I0112942

SCAPEGOAT

What the Invasive Species Story Gets Wrong

Clare Follmann

AK PRESS

Scapegoat: What the Invasive Species Story Gets Wrong

2026 by Clare Follmann

This edition © 2026, AK Press (Chico / Edinburgh)

ISBN 9781849356213
E-ISBN 9781849356220
LCCN: 2025935684

Please contact us to request the latest AK Press distribution catalog, which features books, pamphlets, zines, and stylish apparel published and/or distributed by AK Press. Alternatively, visit our websites for the complete catalog, latest news, and secure ordering.

AK Press
370 Ryan Ave. #100
Chico, CA 95973
www.akpress.org

AK Press UK
33 Tower Street
Edinburgh, Scotland EH6 7BN
akuk.com

"Kinder than Man" by Althea Davis was originally published in *Poems for the Weeping Kind* (2023) and is reproduced with permission of the author.

Cover design by Casandra Johns

Printed in the USA on acid-free paper

This book is dedicated to Pete,
and to the world we have been imagining.

Speech is a spell, and words, once ejected into the air,
warp the weave of worlds.
—Ho Tzu Nyen

We cut nature up, organize it into concepts, and ascribe significances as
we do, largely because we are parties to an agreement to organize it in this
way—an agreement that holds throughout our speech community and is
codified in the patterns of our language.
—Benjamin Whorf

Kinder than Man

And God,
please let the deer
on the highway
get some kind of heaven.
Something with tall soft grass
and sweet reunion.
Let the moths in porch lights
go some place
with a thousand suns,
that taste like sugar
and get swallowed whole.
May the mice
in oil and glue,
have forever dry, warm fur
and full bellies.
If I am killed
for simply living,
let death be kinder
than man.

—Althea Davis

Contents

SCAPEGOAT

The Judas Goats

I have sinned in that I have betrayed the innocent blood.
—Judas Iscariot

Once upon a time. . .

This is traditionally how many stories begin. Although this book isn't a story, it's certainly about a story. It is full of many different stories, and it also begins with a story. So, for the sake of tradition (which this book at times will at length disavow), let it also begin this way.

Once upon a time, there were three goats. . .

The thing about stories is that there is always more than one way of telling them. There may be infinite ways, but in many traditional stories—the ones with a protagonist and an antagonist—there will be two main sides, two different lenses through which to understand the story. I'm sure you have experienced this. You've been caught up in a conflict with another person, and they see the conflict through their lens, and you see it through yours. They think they are right, and you think you are right. Though these lenses may appear contradictory, to understand each side equally is to come to a deeper and more nuanced conclusion about the story as a whole. This allows us to expose the underlying issues embedded in each lens and in each side. By exposing the nuance of a two-sided story, we sometimes find that the source of conflict lies elsewhere.

This book tells the story of invasive species. In one telling of this story, there are two very clear sides. One side contains our protagonist: the human protector of native habitat and biodiversity. The other side, our antagonist, is the enemy invasive species. It is a classic war story, a regular us-versus-them story. In an attempt to explain both perspectives of this story, this book will reveal the true source of conflict that is coming from somewhere beyond these two sides. This book shows that there is more than one way to tell this story.

This book really begins with a story about three goats, but this goat story could not have happened if it weren't also for a group of human fishermen. So let me begin again.

Once upon a time, there was a group of fishermen. . .

It was the summer of 1959, and a group of fishermen set sail to Pinta, the smallest, northernmost island of the Galápagos. The fishermen planned on using this equatorial island as a home base of sorts while they fished for a few weeks, so they brought plenty of essentials to keep them comfortable and well fed. This included three plump goats, which would serve as a backup food source if the fishermen were unable to catch enough fish to sustain themselves.[1]

Whether their fishing expedition was successful or not is unknown, but we can assume that the fishermen's nets were heavy and full for the duration of their trip because these three lucky goats were spared and left to live out the rest of their days on the beautiful island of Pinta. For the goats, this small island proved to be a paradise—so much so that by the 1970s, they had multiplied to over forty thousand, spreading across the twenty-three-square-mile island. But the native plant life of Pinta suffered thanks to these hungry new inhabitants. The goats did what goats are well known to do—like nature's lawnmowers, they spent their lives consuming all the diverse vegetation they could find. Over time, other mariners (like whalers) brought more goats to several of the other Galápagos Islands, and across each island, a similar story unfolded. The goat populations soared while the native plant populations dwindled.[2]

From 1995 to 1997, as the total goat count in the Galápagos topped one hundred thousand, concerned conservationists decided that enough was enough. They needed to put an end to the ecological devastation—and fast. The seemingly endless consumption of vegetation by the goats was also harming the habitat of an endangered native tortoise species. To stop the goats and their insatiable hunger, conservationists initiated Project Isabela.

Spanning nine years—from 1997 until 2006—Project Isabela attempted to restore biodiversity to the Galápagos Islands by "managing" these goats. But the restoration would only be possible if the project achieved complete extermination of the goats. Every single fertile goat across the Galápagos Islands would need to be killed.

"The trickiest part of the whole thing was logistics," said Dr. Karl Campbell, the field manager who ran Project Isabela's operations. "We had to bring helicopters, rifles, and munitions to one of the remotest parts of the world."

The conservationists and volunteers of Project Isabela began by hunting the goats from both the sky and the ground. Exterminators wielding rifles leaned out of roaring helicopters or marched across the tropical landscape, shooting down the goats.

"It can be hard to see so many goats lying dead out there," said Dr. Linda Cayot, science adviser for the Galápagos Conservancy. "But those goats were destroying the habitat of the tortoises. In my heart and mind are the tortoises."[3]

Despite this methodical massacre, the conservationists still faced a problem. While the goat population rapidly dwindled until only a few herds remained, it wasn't enough. Project Isabela required the total eradication of the goats, but the goats were getting wise to the pattern of slaughter, and they learned to hide from their executioners. The conservationists now had to figure out how to find the last remaining goats, which proved far more difficult than originally anticipated.

The masterminds behind Project Isabela came up with a new plan. This involved capturing a selection of the remaining female goats, ominously referred to as the "Judas goats." These Judas goats would then be sterilized so that they could not reproduce, injected with hormones to produce constant arousal, and accessorized with brightly colored radio-tracking collars. The Judas goats were then let loose on the islands—and surreptitiously pursued.

Goats are pack animals; they seek out the company of their own kind. Like us, they don't like to be alone. So the Judas goats roamed the islands, searching for the remaining herds. When a Judas goat at last found her kin, the snipers tracking her would swoop down and massacre the herd, leaving only the Judas goat alive. After watching her companions gunned down around her, the frightened Judas goat would flee and look for more goats, and the bloodshed would continue. At long last, every goat was found.[4] Project Isabela was a tremendous success, and because the sterilized Judas goats could not reproduce, they were allowed to live out the rest of their lives on the islands. This project is an example of invasive species management, albeit a particularly brutal one.

"You have to understand," says Campbell. "We don't want to put a cute face to an invasive species. We focus on the outcomes, letting people see how good the islands look once the goats are gone."[5] This is the strange story of the Judas goat—or rather, the story as told from one side. If told by the goats, the story would be much sadder.

Though the goats were killed off, the story does not quite end there. Several years after the completion of Project Isabela, a study emerged that revealed an unexpected consequence of this goat extermination. The native Galápagos hawk, which had fed on carcasses of the invasive goat species for decades, experienced a drastic decline in population that coincided perfectly with the timeline of Project Isabela. Luckily, the dwindling Galápagos hawk made a few dietary changes and became more reliant on an invasive rat population for food.[6] However, in 2012, conservationists initiated Project Pinzón in an effort to eradicate the invasive rats.[7]

Do you see where this story is headed?

As for the native tortoises, well, this species still fights for survival. As of 2022, it was believed that the decline of native tortoises was partially due to hunting for the sale of their meat.[8] And, despite the elimination of the invasive goats, the tortoise habitat is still in danger thanks to other anthropogenic causes. According to a study published in *Biological Conservation*, if we humans continue our business as usual, these tortoises (among many other species) are likely to become extinct within the next thirty years due to climate-related ecological damage.[9] And it will take more than helicopters and bullets to remedy that source of devastation.

Climate crisis aside, there remains an intricate yet often ignored relationship between anthropogenic interference and the presence of invasive species (with their subsequent ecological damage). It is people who spread invasive species. After all, let us not forget how the goats arrived at the island in the first place. The history of invasive species management is littered with stories similarly dripping with irony. Yet the narrative of invasive species is often only told from one side, which describes them as an invading enemy to be eliminated at any cost. Despite being ethically dubious, the popular ways of "managing" invasive species are typically Sisyphean, doomed from the outset. Invasive species have an impeccable tendency to return to the habitats from which they've been removed—they integrate deeply into novel ecosystems that meet their needs, then spread and multiply. Even if some are killed in invasive species management projects, enough are typically able to survive and to reproduce. The absolute eradication of any invasive species is nearly impossible—except in rare, isolated ecosystems, like islands. Additionally, most common forms of invasive species eradication will also include unforeseen ecological damage. Sometimes, as in the case of the Judas goat and the Galápagos hawk, the invasive species has become so entangled in the intricate web of its new ecosystem that the eradication will disrupt that ecosystem. Eradication methods, like poison or bulldozers, can harm the whole ecosystem, including its native inhabitants. After all, poison and bulldozers cannot tell the difference between an invasive species and a native one. Particularly in the wake of the climate crisis that threatens nearly all of earth's habitats, and with the growing flight of climate refugees, we are facing a strange new world where increasingly outdated stories of *native* habitats, of *native* species, and of Sisyphean invasive species management nevertheless maintain a firm hold on public belief and conservation policy.[10]

Project Isabela was one of the "largest, most ambitious ecosystem restoration project[s] in a protected area worldwide."[11] Yes, much of the flora devoured and infringed upon by these Galápagos goats has since grown back, but, with the looming threat of a global temperature rise of two degrees Celsius (or more) and nothing substantial in the works to protect against it, how long will this already fragile ecosystem survive? Funding ecosystem restoration without any steps to mitigate climate crisis or build a more sustainable way of living ignores the elephant in the room—or perhaps it takes good care of the elephant while allowing the room to burn.

The story of the war against invasive species is often told as a heroic tale of the preservation of endangered species and bio-diversity. Yet an unchecked climate crisis perpetuated by a global capitalist society would only continue to threaten biodiversity, even if all perceived invasive species were to be slain. These conservation efforts are therefore ultimately in vain. This old war story is tired, and it is time to create a new one.

As Greta Thunberg argues, when your tub is overflowing, the first thing to do is turn off the tap. Pursuing invasive species man-agement while largely ignoring climate crisis and ongoing global capitalism is like bailing out the water before turning off the tap. This isn't to say we don't need those sorts of solutions—bailing water is necessary when you have an overflowing tub. But we also must turn off the tap, or our efforts will be pointless. We need far more diverse solutions, or this story and many other stories will simply end in extinction.

I don't deny that the goats on the small Galápagos Islands caused great ecological damage and that they threatened already endangered species that deserve protection. Yet there can be no doubt that Project Isabela offers an extremely gruesome and contro-versial way of eradicating a species. In the story of the Judas goat, I can't help but feel a deep compassion for those unlucky goats, mer-cilessly hunted down in their new home for the sake of imprudent conservation efforts. But this is largely the way that scientists in invasive species management have decided to address the problem of biodiversity loss. It is just this sort of slaughter that is generally considered necessary for the greater good of the ecosystem. It goes largely unquestioned.

But I believe it is important to ask questions. I believe it is espe-cially important to ask questions when something feels wrong but is made to appear unquestionable through language, rhetoric, or tradition.

To what extent is invasive species management necessary? Where do the moral or ethical limitations lie in this philosophy of preservation? What has brought invasive species management to this degree of brutality, and what maintains it? How can we protect biodiversity in a way that feels right? How can we create more ethical, sustainable, and effective forms of invasive species management that do not themselves cause harm to the ecosystem we are trying to save? How can we live with nature better?

In this book, I first break apart and show how the current story of invasive species management works. That is to say, what words, what imagery, what symbolism are embedded in the narrative that describes the war we wage on invasive species, and why are they so effective? It is a persuasive and misleading story—*how* are we being persuaded, *how* are we being misled? And then I go back further to explore science's historical relationship with language and communication with the public—or, put in another way, science's relationship with storytelling.[12] I then expose the political, ethical, cultural, and scientific holes in the current invasive species story. I argue that this story is told so vividly and so viciously because it serves as a distraction from the story's true source of conflict (the tub's gushing tap): a capitalist way of living that perpetuates climate crisis. After unraveling this story, I offer new threads from which we might form a new one.

This story, like so many, is deeply entangled in our contemporary capitalist culture. And it is clear that this capitalist way in which we are made to live—reliant on the exploitation of both nature and people—is fundamentally opposed to an ethical and sustainable way of living with the earth and with each other. As philosopher Fredric Jameson famously declared, it is easier to imagine the end of the world than the end of capitalism. However, I seek to challenge this convention and spark a more inspired imagination. By rewriting this story of invasive species and conservation under capitalism and climate crisis, I am hopeful that we can find new ways of existing with and relating to both other people and our natural habitats—ones that are anticapitalist and non-extractive in nature.

But first we must unravel and question some stories we have been told.

We earthlings stand together on the precipice of a catastrophic anthropogenic climate crisis.[13] Particularly with the emergence of climate refugees, the current popular framework of invasive spe-cies management is obsolete—it is dead. But this narrative remains propped up by modern ecology and made to seem alive and well—think of *Weekend at Bernie's*.[14] When such an outdated perspective is made to withstand its own extinction, we must ask ourselves why.

In my writing, I have found it difficult to come up with a word to succinctly describe the particular way we are made to live with (or, in this case, against) nature and each other. This struggle indicates a larger difficulty of defining this era in which we find ourselves. How best to define the global social, political, cultural, economic present? Anthropocentric? Colonial? Capitalist? Eternally extractive? Consumptive? Cancerous? Technophilic? Western? Suicidal? You will see I have given up and resorted to the catchall term "capitalism" because most of us will likely know what I mean when I use this word, and it generally encompasses these qualities. But I'd like to briefly explain why the term falls short.

Capitalism requires individuals to organize their ability to live (or live well) based on what they own—their property, the value of which is largely based on societal beliefs. Capitalism functions according to individuals' ability to make a profit off their property as well as their ability to acquire more property. This creates a grow-or-die system of competition that exploits both humans and the natural world, one in which the wealthy control the economy. Capitalists spend money to acquire more money through the form of labor: hiring workers to transform their property into some-thing of even more value, then making a profit according to how little they can pay their workers for this labor. Capital and the state work hand in hand to ensure that a society with a vast pool of such workers exists, privileging some workers according to race, gender, class, and ability in order to weaken cooperation with each other and opposition to this system. The effect is to force the majority of the people on earth into destitution, with no other way to survive besides selling their labor.

In a society built on exponential growth, our relationship to our world can only be extractive in nature. On a finite planet, eternal extraction in pursuit of profit can only lead to a hollowed-out, dead world—a world that has been exhausted of all its profitable offerings.

Food, water, air, land—that is to say, *life*—are all absorbed into the blind consumption machine that is capitalism. It is this definition, and this simple and logical conclusion—this natural catastrophe of capitalism—that I am implying when I speak of capitalism throughout this book. Capitalism is a death sentence for our world. We are part of our world. Capitalism is a death sentence for us.

But this extractive tendency extends beyond and even predates the rise of capitalism. Imperial states, monarchies, hierarchical groups, communist countries, colonizing societies—these nations and ways of living will all likely bow before exploitation and extraction.[15] In light of this, the term "capitalism" describes what is just a much more efficient (and that much more destructive) mechanism for a tendency—hoarded accumulation and unfettered consumption—that has been manifest in other moments of human history. However, in our current global society and under the ongoing climate crisis, capitalism is without a doubt our largest barrier to a better life. I hope that in its wake we will be able to build a social structure that is not so mortally exploitative.

At any rate, I am using what I believe to be the best term I have available: "capitalism." I hope this explanation clarifies or at least provides some context for the nuance embedded in the semantic riddles that follow.

CHAPTER 1

The Nature of Language

Why the Story of Invasive Species Management Is So Effective

Words are not bound directly to other pictorial elements.
They are merely inscriptions on blobs and shapes.
—Michel Foucault

There is extensive discussion on the topic of invasive species among researchers and practitioners in fields relating to land management. The general consensus among many professionals as well as nature enthusiasts and backyard gardeners is that invasive species are unequivocally bad and that it is our collective duty to exterminate them lest the ecosystem in which they thrive become worthless. The disinclination to question this apparent truism is owed, in large part, to the language and rhetoric of invasive species management. It is through the combined power of metaphor, jargon, and a little-known class of persuasive terms called "plastic words" that a tempting tale is spun of the irrefutable evil of invasive species. But to understand the full picture of this linguistically deceptive story, it is important to take a step back and unpack how words work.

A Brief Overview of Language and Semiotics

Semiotics is "the science of communication studied through the interpretation of signs and symbols as they operate in various fields, esp[ecially] language."[1] There were two key figures in semiotic theory at the turn of the twentieth century: Ferdinand de Saussure and Charles Peirce. Saussure's work is foundational for semiotic linguistics, and Peirce elaborated on Saussure's ideas. Semiotics, at its core, describes the relationship between the *sign*, the *signifier*, and the *signified*. According to Saussure, the signifier is an indicator that is understood to represent the signified, and the sign is the name given to this relationship relationship between the signifier and the signified.

For example, the word "pipe" (or a photograph, painting, or drawing of a pipe) represents the concept of what a pipe is. The word (or artful rendering) is the *signifier*, and the concept of a pipe is the *signified*. Meanwhile, the *sign* refers to the relationship between the word or image (signifier) and the concept (signified). A signifier doesn't have to be a word or a depiction. For example, another signifier for a pipe could be encountering the syrupy, pungent smell unique to pipe tobacco. Each of these signifiers does not refer to any specific pipe but rather represents the general idea of pipes.

Peirce elaborated on these distinctive signifiers (the word "pipe" vs. smoke vs. the image of a pipe) by noticing that while they may signify the same thing, they each have a distinctive nature. He argues that there are then three types of signifiers: *icon*, *index*, and *symbol*. An *icon* is a close visual depiction of the signified, for example, that painting or photograph of a pipe. An *index* is an experiential phenomenon that is concurrent to or caused by the signified, as in the smell of the smoke the tobacco produces when burned in a pipe. A *symbol* is only able to work as a signifier because of widespread agreement or convention that it will signify the signified, for example, the word "pipe."[2]

The word "pipe"—which has no inherent relationship to the object—is nevertheless still recognizable (at least by those who speak English) as signifying a pipe. Both icon and symbol are arbitrary, meaning they do not have a concrete, literal connection to the signified. We can recognize both the icon and the symbol as signifiers only because our culture connects these signifiers to the signified. For example, a cat may look at a painting of a pipe but will probably not make any connection between the painting and a physical, material pipe. Similarly, most cats will not be able to understand the word "pipe"—except in the course of repetition and direct association, as when they learn to understand what "treat" means just before they are given one.

However, a cat can smell pipe tobacco and know that the source of the smell, the pipe, is somewhere present. The index is not arbitrary. Semiotician Roland Barthes argues that words have multiple dimensions: as "a cultural reference, a rhetorical model, a deliberately ambiguous utterance, and a simple indicative unit."[3] In short, words are not what they represent. They are the *signifier* (or the *symbol*) that *signifies* the *signified*, just as the word "rose" suggests the fragrant, many-petaled, dew-touched, sharp-stemmed, historically, literarily, and culturally charged romantic flower. Semiotics tells us that language is a system of meaning. There is no objective meaning—only a nuanced and forged collective understanding in the many-layered context of language, which will be very particular to that language's culture. Yet this is how we fundamentally understand the world: by means of conceptual frameworks mediated and organized through language. Language is key to how we make sense—how we create comprehension. Words mean nothing in and of themselves but everything to us because words structure how we perceive the world.

Consider the artist René Magritte's famous painting *This Is Not a Pipe* as well as the essay of the same name by philosopher Michel Foucault. James Harkness's introduction to his translation of Foucault's essay explains:

> In Saussurean linguistics, words do not "refer" to things themselves. Rather, they have meaning as points within the entire system that is a language—a system, further, conceived as a network of graded differences. "Dog" is not somehow attached to the real animal, arising naturally from it and participating magically in its essence or presence. Instead, "dog" has conceptual signification insofar as it evokes an idea that differs from the idea of a cat, a bear, a fur seal, etc. It has syntactical signification insofar as it (a noun) differs from words such as "bark" (verb) or "furry" (adjective) and thus cannot take their places in a proposition; and it has phonetic signification insofar as it differs from more or less similar sounding signifiers such as "bog," "dot," "dig," and so on. From the commonsense vantage this seems an unnecessarily complex and circumlocutory

approach to language, aimed at the most radical divorce possible between words and things. And why bother? After all, would anyone seriously argue that a word is what it represents—that the painting of a pipe is the pipe itself? . . . Yet it is exactly from the commonsense vantage that, when asked to identify the painting, we reply "It's a pipe,"—words we shall choke on the moment we try to light up.[4]

Here Harkness elaborates on how arbitrary symbols and icons really are. Words and language are symbols and can only operate successfully within a complex system of meaning, history, shared understanding—that is, an entire culture. Because words work within language, and language works within culture, words themselves carry the weight of culture, society, story, and history. This is why it is so often difficult to remember that these symbols and icons are arbitrary.

But when we remember this, we are made to recognize how incredibly powerful words are, able to convince us that this arbitrary, phantom connection is real.[5] We are likewise made to recognize how suspicious we must be when some words (particularly heavily charged words) are used in lieu of other words to describe things, as in the form of both metaphors and plastic words.

The persuasive and misleading story of invasive species and invasive species management relies heavily on metaphors as well as plastic words. We must be suspicious of this reliance, because how we talk about things can hide or embed ideologies in our ideas of those very things—we can sometimes even forget the subjectivity of rhetoric. But first, what are metaphors, and what are plastic words?

Metaphors

"Metaphor" is defined in the *Oxford English Dictionary* as 1) "a figure of speech in which a name or descriptive word or phrase is transferred to an object or action different from, but analogous to, that to which it is literally applicable," and 2) "something regarded as representative or suggestive of something else, esp[ecially] as a material emblem of an abstract quality, condition, notion, etc.; a symbol, a token. Frequently with *for*, *of*." The etymology of the word originates from the Greek *metepherein*, meaning "to transfer," where *meta* means "over" or "across," and *pherein* means "to carry or bear."[6]

Put differently, we can understand metaphor as a rhetorical device that marries two different ideas from two different realms, collapsing them into one notion. To use a metaphor to describe metaphor: metaphors are linguistic alchemy, transmuting one thing into something completely different. Where a simile will say one thing is *like* another, a metaphor will say one thing *is* the other. In poetry, literature, or drama, the use of metaphors tends to be stylistic, invoking aesthetic elements of the world around us that stir emotion.[7] In William Shakespeare's *Romeo and Juliet*, Romeo cries, "But soft, what light through yonder window breaks? / It is the east and Juliet is the sun."[8] Metaphors are effective tools of persuasion because they transfer emotion from something familiar or universal, like the brilliant, warming sun, into the specific something that the speaker wants the audience to understand, like Romeo's unique experience of his love for Juliet.[9] Metaphors can help an audience to empathize, to feel the emotion of another.

You may have noticed that I rely on metaphors in my writing—a stylistic choice to help me paint a vivid picture, since much of what I'm writing about, at least in this chapter, is theoretical. It's true too that I have an argument in this book, and I am writing to persuade. However, I am not claiming to be an expert or authority, and I certainly don't have the power to make policies based on my beliefs laid out here. In the arts—literature, drama, poetry—the persuasive use of metaphor is essentially a harmless stylistic element that serves to illustrate and generate empathy and beauty. But when metaphors are used in disciplines of power—in fields such as policy, government, and science—this persuasive rhetorical tool can be manipulative. As Colin Koopman describes in his article "The Algorithm and the Watchtower": "To make sense of the new political power that can be built out of all this data requires metaphors, and these metaphors themselves are not without political stakes. Metaphors help shape the meanings of the activities in which we are engaged, and they thereby help condition what possible actions we can conceive ourselves as undertaking. Metaphors thus have a political stake in that they define the forms of power that control us and the forms of possible resistance to power we can imagine."[10]

Metaphor, then, can become a tool of propaganda in the hands of the powerful. Some metaphors are obvious, such as "Time is money." Even though this is a bit of poetry, it nevertheless governs how we think about time. The danger of metaphor is when we forget their literary foundation. Essayist Eula Biss writes about anthropologist Emily Martin, who asked a group of scientists about the reliance on war metaphors to describe the immune system. Some of these scientists disagreed that war was a metaphor for the body at work. They argued that this was just "how it is."[11]

In invasive species management, we are similarly force-fed a highly charged metaphor that paints designated species as opposing enemies in an imaginary, metaphorical war. The linguistic alchemy is so thorough that somewhere down the line, this relentless rhetoric might have us really believing that a dandelion is an invading enemy while we spray poison across our lawns.

In short, metaphors are rhetorical devices that simplify complex issues by comparing those issues to more familiar topics. Like a magic spell, they linguistically turn one thing into something else. Metaphors can covertly organize our thought and discourse. Metaphors can, in turn, become sophistic rhetoric: plausible but absolutely misleading. Metaphors can work to manipulate how we think and make us believe that the connection they build between two unrelated things is real.

Why Are Metaphors Used in Science?

In *Metaphors We Live By*, philosophers George Lakoff and Mark Johnson describe various metaphors that have slipped into the vernacular and gone unnoticed. For instance, they describe the conventional understanding that an argument is a battle by calling to mind various phrases such as these:

> Your claims are *indefensible.*
> He *attacked every weak point* in my argument.
> His criticisms were *right on target.*

I *demolished* his argument.
I've never *won* an argument with him.
You disagree? Okay, *shoot.*
If you use that *strategy*, he'll *wipe you out.*
He *shot down* all of my arguments.[12]

Science is no exception. Professor of rhetoric Alan Gross writes, "Science is full of metaphor, and it is the nature of metaphor deliberately to misname."[13] This phrasing implies intentional manipulation, but metaphors are often used in science specifically because they make more complex concepts easier to understand. That is, conveying abstract scientific findings in a clear, comprehensive way often benefits from the use of metaphor. Still, a metaphor is a rhetorical device that sacrifices nuance for simplicity. "The price of metaphor is eternal vigilance," cyberneticists Arturo Rosenblueth and Norbert Wiener explain, and if we aren't careful, through repetition that metaphor might become hypnotic sophistry.[14]

Some Common Science Metaphors

Not all science metaphors are made equal. Scientist Richard Lewontin describes a metaphor called "the billiard ball model of molecules," which is used in physics to explain how molecules bump into and rebound off each other.[15] This metaphor, he argues, is harmless because we use the idea of billiards to make sense of molecular behavior, but we don't actually imagine that molecules *are* little colored balls with numbers on them. In contrast, there are trickier metaphors.[16] For example, health and medicine rely on militaristic metaphors: germs *invade the body*; the patient *fights* or *combats* illness. Invasive species management relies on a similar metaphor of militarism and war. These militant metaphors are easy to comprehend, evocative, and viral. And they pose a threat in their power to mislead. Writer Susan Sontag argues:

> Indeed, the transformation of war-making into an occasion for mass ideological mobilization has made the notion of war useful as a metaphor for all sorts of ameliorative campaigns

whose goals are cast as the defeat of an "enemy.". . . Abuse of
the military metaphor may be inevitable in a capitalist society,
a society that increasingly restricts the scope and credibility
of appeals to ethical principle, in which it is thought foolish
not to subject one's actions to the calculus of self-interest
and profitability. War-making is one of the few activities
that people are not supposed to view "realistically"; that is,
with an eye to expense and practical outcome. In all-out war,
expenditure is all-out, unprudent—war being defined as an
emergency in which no sacrifice is excessive.[17]

Sontag here unites capitalism and war to explain why militant
metaphors are especially successful in shaping our perceptions
and responses. It is because they feed off a common self-serving,
unethical predisposition, which, Sontag argues, we can attribute to
capitalist society. Capitalism praises self-centered survival above
all else, and similarly, in war, it is every man for himself. Sontag
suggests that in capitalist society, we are waging war with everyone
else, simply because they are not us.

Philosopher Franco "Bifo" Berardi suggests, "Only an act of language escaping the technical automatisms of financial capitalism will
make possible the emergence of a new life form."[18] So it is not the
metaphors of militarism but rather, ultimately, the modes of thought
associated with capitalism that may be the true enemy, starving out
alternate perspectives. Fredric Jameson's famous maxim—"It seems
to be easier for us today to imagine the thoroughgoing deterioration
of the earth and of nature than the breakdown of late capitalism"—
unpacks how capitalism possesses a totalizing ability to leech us of
any imagination that would allow us to come up with other ways of
thinking, living, or being.[19]

What happens when wheat becomes a *product* and soil becomes
a *resource*? When these commercial terms are metaphorically applied
to natural phenomena, they change our understanding of the world.[20]
Economic metaphors are rampant in agriculture; is it not because
land today has become a commodity? Food costs money, and so
food has *become* money—at least, metaphorically—consider "bread"
or "dough." In this case, agriculture employs economic metaphors

that can help maintain that capitalist perspective of the land, and the exchange of terminology goes both ways—we refer to stocks or money *growing* in our bank accounts.

Wherever this occurs, in medicine, the farm, or the stock market, we may have ceased to recognize metaphor at work. Once upon a time, wheat was itself, a golden-yellow plant growing from the ground, rippling in the wind. Today, wheat is a *product* grown out of a *resource*—money grown out of money. Metaphors of capital run rampant.

First, food was grown and eaten: food was food. Then food was sold for money, and now food *is* money. This is the great linguistic alchemy of metaphor: money doesn't grow on trees because, well, trees *are* money. These sorts of metaphors are familiar; they fit snugly into place. In these cases, as geneticist Richard Lewontin says, "It's easy to forget that metaphors are metaphors, that they are provisional and limited comparisons, not literal descriptions."[21] These metaphors are strong in part because they are woven through with scientific-sounding words. But when we forget that these are metaphors, we may regrettably think that we know they are *facts*.

Lewontin describes the problem of uniting previously unrelated theories across disciplines, as, for example, the idea of evolution as it is used in different disciplines or arenas. The way human societies change over time could be considered an "evolution," a bridging of ideas that helps to convey an interpretation of this history, but it is still only an interpretation. This interpretation is easy to understand because it adopts the familiar concept of evolution. However, the metaphor of evolution here is essentially an inaccurate application because human societies change in ways that are fundamentally different from Darwin's theory of evolution. Evolution suggests a step-by-step path of increasing one's ability to survive. Societies do not evolve in the way that species do, by becoming better adapted under the pressure of survival, with the reproduction of the predecessors' best-suited qualities.[22] As the philosopher Aristotle warns in *Rhetoric*, "In using metaphors to give names to nameless things, we must draw them not from remote but from kindred and similar things, so that the kinship is clearly perceived as soon as the words are said."[23] We must be careful when we use metaphors. There has always been danger in stretching the metaphor too far.

Of course, language itself is metaphor; words are not the things they describe. You need only repeat a single word out loud a handful of times to have the dizzying experience of understanding of how easily a word can be rendered into gibberish! When calling a chair "chair," we are calling a specific object a name by which many other distinct objects are called. What I am currently sitting on at my desk as I write this book is not by necessity called a chair; "chair" is just one name we have invented to exchange the idea of a chair with each other. My name is Clare, but Clare is not what I am. To borrow again from *Romeo and Juliet*:

'Tis but thy name that is my enemy.
Thou art thyself, though not a Montague.
What's Montague? It is not hand, nor foot,
Nor arm, nor face, nor any other part
Belonging to a man. O, be some other name!
What's in a name? That which we call a rose
By any other word would smell as sweet.
So Romeo would, were he not Romeo called.[24]

Strange that a passage from Shakespeare would provide such strong caution about the power of language to seduce and persuade. Scientists and Shakespeare both toil in the world of metaphor, a medium of linguistic exchange where nothing is innocent.

Plastic Words

In 1988, German linguist and philosopher Uwe Poerksen wrote *Plastic Words: The Tyranny of a Modular Language* and identified a list of thirty or so words that share a remarkable set of characteristics. These "plastic words," so named by Poerksen for their malleability, originate in the vernacular, migrate into scientific discourse, and then return to the vernacular. Over the course of this migration, meaning is lost. In the absence of meaning, these words carry instead a powerful aura of correctness that invites compliance.[25]

Plastic words, Poerksen writes, "have been fashioned for the purpose of laying down the tracks and outlining the routes of a civilization that is covering the globe with gathering speed. Their origins can no longer be discerned. They resemble one another. . . . They may not be noticed, but they are present everywhere. . . . They invade private conversation. When they first appear, they are fashionable and command attention; but they merge with the everyday and soon seem commonsense."[26]

Here is Uwe Poerksen's list of plastic words:[27]

accomplishment	growth	quality
basic needs	health	raw material
capitalization	identity	relationship
care	information	resource
center	living standard	role
communication	management	service
consumption	modernization	sexuality
contact	model	solution
decision	partner	strategy
development	planning	structure
education	problem	substance
energy	process	system
exchange	production	value
factor	productivity	work
function	progress	workplace
future	project	

I'm certain they sound familiar to you. But what do these words mean? Resources—are they human resources or natural resources? Do we know if they are growing or shrinking? Are we supposed to put money into these, or are they money already?

This is how plastic words show up, how they dig their heels in and comfortably make themselves at home. Their persuasive effect is contingent on their pervasiveness combined with the plasticity of their meaning. As they circulate, the meanings of these plastic words change. They are flexible enough to be used in a variety of disciplines and can refer to a wide range of concepts. Don't these sorts of words bear a strong resemblance to the rhetorical device of metaphor? They can mean many things—or nothing—at once, and whoever uses them has the power to decide what they will mean. (Sometimes, these words can even mean opposite things when used in different contexts—more on that later.) Because plastic words are overarching, they make more specific terms obsolete. They are difficult to argue against because they are so familiar. Most people have a general idea of their meaning, but it can be hard to know exactly what is being said when plastic words are being used. They can render opposing points of view ineffective because they cast such a broad net, eliminate specificity, and feel official. "The word signals science. It silences," Poerksen writes.[28] With their variability, ahistorical quality, and their silencing effect, plastic words help to make inaccessible whatever topic they permeate.

Language is alive and, like an organism, evolves over time. As philosopher Ludwig Wittgenstein writes, "The meaning of a word is its use in the language."[29] Meaning does not precede use. With changing usage and context, certain words can become more plastic, while other formerly plastic words can lose their plasticity. Uwe Poerksen admits that his list is not comprehensive, but I think you will agree that there is a certain aura to these words, and a feeling that they seem somehow authoritative.

Poerksen notes thirty characteristics of plastic words to show the ways they simplify, reduce, and homogenize language, decreasing its precision and contextual efficacy. These are their characteristics paraphrased:

1. The speaker lacks the power to define the word.
2. The word is superficially related to scientific terms. It is a stereotype.
3. It has its origin in science.
4. It is carried over from one sphere into another and is in that sense a metaphor.
5. It forms an unnoticed link between science and the everyday.
6. It has a very broad application (domain of use).
7. It displaces other possible, more clarifying synonyms.
8. It replaces the conventional, precise word.
9. It substitutes for an indirect way of speaking or a silence.
10. It condenses a huge field of experience into one expression.
11. It is impoverished in content.
12. Its imagery is vapid and diffuse.
13. It is historically disembedded.
14. It transforms history into a laboratory.
15. It dispenses with the question of value.
16. The aura and associations of the word dominate over its meaning.
17. It names a property and gives the appearance of an insight.
18. It has more function than content.
19. As a scientific idealization of something limitless, it uncovers and awakens needs.
20. Its apparent naturalness strengthens this pull.
21. The resonance of the word is imperative.
22. It has multiple uses.
23. Its use increases prestige.
24. It leads to silence.
25. It anchors the need for expert help in the vernacular, and it serves as a resource.
26. It forms new words and is a flexible instrument in the hands of experts.
27. It makes pre-existing words look out of date.
28. In this sense, it is new.
29. It is an element of an international code.
30. It lacks intonation and cannot be replaced by pantomime or gesture.[30]

Plastic words are the siblings of scientific jargon but not twins. As adoptees of science, they neatly bridge the pseudo-objective world of science with the everyday, but they do so discreetly. They carry the weighted power of science but are in fact lightweight in meaning and signification. Weightlessly, they can be easily transported across radically different concepts, realms, or disciplines and still promenade a sense of scientific power. They erase history and context, replacing a more precise and accurate explanation with a solitary empty word. Because they flow from science back into the vernacular, plastic words radiate a power that is far more important than anything they might actually mean or suggest—they are bloated with scientific authority but ultimately hollow. They effect a camouflaged confusion; they lack consistent meaning from context to context yet are often used by experts or officials in their explanations, making it appear as if one must always rely upon experts in power to know and understand what is being said. They serve to evoke authority rather than to offer clarity.

These plastic words are designed to silence the recipients of the message: they do not offer room for contestation, conversation, disagreement, or alternatives as they are all-encompassing in their vapidity. The audience is intended to do nothing but receive, absorb, and obey. In 1966, Situationist Mustapha Khayati described this silencing tendency of bureaucratic language: "People no longer even need to talk to each other: their first duty is to play their role as receivers in the network of informationist communication to which the whole society is reduced, receivers of orders they must carry out."[31] This silent audience is a symptom of an industrial language that is certainly not new, and things have only gotten worse.

When used by institutions that hold power over our lives, plastic words can affect an uncontested, almost tyrannical influence. They slip under the radar. As with metaphors, our familiarity with plastic words makes it hard to notice them. When we use or repeat plastic words thoughtlessly, we degrade our own ability to be critical or self-reflective. The sum effect works against accountability and responsibility.

Poerksen argues that plastic words so easily dominate because "our language is quite fragile, has little native resilience."[32] Is it worth noting how plastic words strike more than a passing metaphorical resemblance to the way invasive species are described? Crossing borders or contexts, eradicating other, weaker competitors, replacing diversity with singularity. This quick ability to adapt, Poerksen notes, is why plastic words are "so well suited for colonization."[33]

Consider "development," for example, which is a plastic word that seeps easily into many contexts. A parent is supposed to track their child's sequential achievements to establish a pattern of "normal" child development.[34] A fetus physically develops in similar step-by-step degrees. The building of apartment complexes is also considered development. Film develops. We can recognize the truth in these different uses of the word. Development is the movement, growth, or act of becoming something bigger, better, transforming into something desirable. Yet the original definition of "develop" comes from the French *developer* circa the twelfth century, meaning "to free (a person from something), to unwrap (something), to unfurl, open out (something)."[35] Language itself is a fluid entity. Like a river, it moves and changes and routinely rewrites its course in gradual increments. The development of the word "development" from its twelfth-century meaning to our contemporary understanding is not necessarily sinister. But what makes "development" and all plastic words so treacherous is their camouflage, and how they are so easily able to justify and disguise destruction.

"Development" has many definitions, but above all it has the taste of something good that is in the works. When something develops, we are made to believe it becomes better, more valuable, more usable. This, however, neglects a darker history of the word. We would do well to ask: What is the fate of the undeveloped or underdeveloped?

On January 20, 1945, Harry Truman coined the word "underdeveloped" in his inaugural presidential speech to describe certain regions that together make up more than half the world. Philosopher Wolfgang Sachs writes, "For the first time, the new world view was announced: all peoples of the earth were to move along the same track and aspire to only one goal—development."[36] In this instant, a new way of seeing the world was born, and the world's population set suddenly on the course toward the goal of becoming developed.

Under the banner of development, the United States legitimizes invasion and intervention of other countries and cultures. In George W. Bush's military campaign Operation Iraqi Freedom in 2003, "some 140,000 U.S. troops [were] deployed in Iraq, in addition to civilian experts and US contractors, who provide[d] substantial support to their Iraqi counterparts in the fields of security, governance, and *development*."[37] The total number of Iraqi civilian deaths by violence from the beginning of Operation Iraqi Freedom through 2020 is estimated at between 187,499 and 211.046.[38] Here, "development" means massacre. As Poerksen writes, "With a word such as development, one can ruin an entire region."[39]

This nebulous quality of words has been identified time and again. Michel Foucault writes, "A word should be taken in its most literal etymological sense; in such a state, things are 'laid,' 'placed,' 'arranged' in sites so very different from one another that it is impossible to find a common place beneath them all."[40] Like Richard Lewontin's critique of incongruous metaphors, Foucault argues that uniting certain separate ideas is inappropriate.

Plastic words create an amorphous, chameleon-like, zombie language. They are the graffitied walls of the Tower of Babel, the language of the Leviathan, the native tongue of the machine. Plastic words serve as industrial capitalism's armor and are used to justify almost any action, even as this results in the abuse of people and the planet. Through a reliance on plastic words that bear a false mantle of science, highly subjective cultural, ethical, and profit-motivated agendas are snuck in through the back door.

Plastic Words, Metaphor, and Semiotics

As I have said, the function of plastic words is akin to that of metaphors. They both carry a specific idea from one domain to another, completely different domain.[41] Plastic words thus share a similar danger to that of the overused metaphor conventionally misapplied. As Poerksen describes it, "The chasm between the sphere of origin and the sphere of application is easily overlooked. There is no tension; no spark jumps between the two spheres. They are tied together seamlessly. Their original separation is hardly remembered.

The result: one takes the word for the thing."[42] As with metaphors, when used so integrally in a new context, the original meaning of the words is lost, but most importantly this occurs inconspicuously. It is as if these terms have always been there in this other context, sphere, or discipline.

In a case of plastic word as metaphor, writer Eula Biss explores the concept of an immune system in a series of interviews, revealing an instance in which metaphor and plastic word appear together under the banner of science:

> The term *immune system*... was probably a metaphor from its very introduction. In a medical context, the word *system* traditionally referred to a collection of organs or tissue, but the immunologists who first adopted it were using it in a broader sense. "Why was the term immune system accepted so widely and so rapidly?" asks the historian of immunology Anne Marie Moulin. The answer, she suggests, resides in its "linguistic versatility," the ability of the term to contain many concepts and multiple understandings. It entered the mainstream just a few years after its introduction to science, spilling into the popular usage of the 1970s. "Though the term was borrowed from the science of immunology," Fitzpatrick writes, "its new meaning was filled out with ideas derived from influential contemporary trends, notably environmentalism, alternative health and New Age mysticism."[43]

Here we see how the metaphorical phrase "immune system" helped us forget it was a metaphor and had us take it for fact. The phrase itself was stripped of its original meaning and infused with a scientism that allowed it to carry many new and different meanings, like our plastic words. Again, this is not to say that a gradual change of meaning that reflects societal and cultural change renders language meaningless—far from it. The ability for language to warp and change alongside its culture is a testament to its aliveness. Words are variable, mutable, interpretable. It is a far graver problem when a word can change in meaning at a whim or when context provides no illumination of a word's meaning, as is quite often the case with plastic words. Plastic words mimic the aliveness of language in their mutability but have no story and will differ from context to context, mouth to mouth. A plastic word shares little to no unanimous consensus of definition. The power of semiotics is visibly at work in both the concepts of metaphor and of plastic words.

In semiotics, the random *signifier symbol* (the word) and the specific *signified* (the object or concept) are united. Two totally separate things, united by consensus. Metaphors serve a similar purpose of marrying differing concepts together, although perhaps with a little less consensus. Poet Anne Carson describes metaphors and how they resemble the theory of semiotics in her book *Eros the Bittersweet: An Essay*:

> There is in the mind a change or shift of distance, which Aristotle calls an epiphora (Poet. 21.1457b7), bringing two heterogeneous things close to reveal their kinship. The innovation of metaphor occurs in this shift of distance from far to near, and it is effected by imagination. A virtuoso act of imagination brings the two things together, sees their incongruence, then sees also a new congruence, meanwhile continuing to recognize the previous incongruence through the new congruence. Both the ordinary, literal sense and a novel sense are present at once in the words of a metaphor; both the ordinary, descriptive reference and a novel reference are held in tension by the metaphor's way of looking at the world.[44]

Metaphors construct new semantic relationships between disparate things. Metaphors bridge differences. Similarly, semiotics tells us we are uniting a thing—whether that which we sit on, often accompanied by a flat object of similar height (*chair*, *table*); the emotion we feel when we are near someone we care so very much for (*love*); that which we delicately pluck, give to that someone we care very much for, might stop and smell a moment amid our busy days, and onto which bees alight (*a rose*)—with a word made of letters, themselves made up of spaces, lines, dashes, and dots, that we decide will be that something's name.

Similarly, plastic words unite one unrelated term (the plastic word) with many different concepts. Uwe Poerksen explains the ways in which plastic words *signify* and the ways they *mean*: "So, do 'development' and 'sexuality' mean the same thing? It seems to me that they signify different things, but what they signify is less important than what they mean. And the meaning is the same. These are close relatives of the myths of everyday life described by Roland Barthes. They are idols, magical and empty."[45] While these words may indeed *signify* different concepts ("sexuality," of course, does not signify the same thing as "communication," "development," or "plan"), what they *mean* is everything and nothing at the same time. What they mean is *science*. They mean *authority*. They mean "Believe this." Here Poerksen's wordplay stresses the semiotic specificity of signification (what a word represents) as opposed to the ambiguity of what something *means,* in terms of both sense-making (what does it mean?) and judging its significance or importance (what does it mean?). Rather than create new concepts, plastic words seem to diminish our individual and societal understanding and perception, failing to call our attention to any appreciable meaning-making at all, instead acting as stand-ins or placeholders.

There is one realm in which plastic words and the principles of semiotics diverge: in translation. You can say *rose* in any language, but it will never carry over absolutely the nuanced, star-crossed Romeo-and-Juliet cultural meanings into another. Plastic words, in contrast, can carry over almost seamlessly from language to language, with nearly infinite references inside the system of meaning. Because they are so hollow in significance, their vague, evocative associations differ from those of words that are set to function with specificity in a given system of meaning, of language and its culture. For example, Poerksen describes a conference in Mexico which "discussion was dominated by a number of words that floated through it like driftwood: *'progreso,' 'proceso,' 'modernizacion,' 'necesidades,' 'comunicacion,' 'informacion,' 'crisis,' 'desarollo.'*"[46] Though masked slightly by another tongue, these words stayed largely the same from language to language (after all, Poerksen's text itself is translated from German, yet his word list remains just as relevant in English).

Chillingly, plastic words are so devoid of meaning that they are symbols that can mean anything.

When plastic words and persuasive metaphors together tell a story, an audience can be made to wholly believe an oversimplification. This can have profound consequences.

Invasive Species Management

The very phrase "invasive species management" is made up of a militaristic metaphor and a plastic word. Both the scientific and the public discourses on the topic of invasive species are dangerously full of inaccessible language and sophistic rhetoric.

The metaphor in "invasive" conjures militaristic ideas of an invading other, an enemy, or war. Biologist Charles Elton coined the term in *The Ecology of Invasions by Animals and Plants* to describe nonnative species that harm the native regions into which they migrate. Elton wrote his treatise in a postwar United States, and the text is filled with military terminology.

"Management" shows up in Poerksen's list of thirty plastic words. A marriage of the prefix *manage* and the suffix *-ment*, it comes from the Italian *maneggiare*, from *mano* meaning hand, which derives from the Latin *manus*. *Maneggiare*, when used in the mid-sixteenth century, meant to handle—specifically to handle or train a horse. Related is the Spanish *manejar*, meaning to use or manipulate. Other early uses of the word "management" implied manipulation or trickery.[47]

Today the meaning of the word is up for interpretation. Far removed from these original definitions, the management of workers is today an obscure role that very rarely includes the handling of horses, and likewise most managers will be quick to assure you that manipulation and trickery are as far from their list of duties as horse training is. In invasive species management, the word means something more like "control" or "diminishment" or "eradication"—say, the wholesale massacre of certain plant or animal species. However, sometimes its meaning even in this ecological milieu is not so clear. For example, an executive order signed by President Barack Obama in 2016, titled "Safeguarding the Nation from the Impacts of Invasive

Species," describes a "management plan" that encompasses many different goals, including to:

> (1) provide institutional leadership and priority setting; (2) achieve effective interagency coordination and cost-efficiency; (3) raise awareness and motivate action, including through the promotion of appropriate transparency, community-level consultation, and stakeholder outreach concerning the benefits and risks to human, animal, or plant health when controlling or eradicating an invasive species; (4) remove institutional and policy barriers; (5) assess and strengthen capacities; and (6) foster scientific, technical, and programmatic innovation.[48]

While some ideas of what "management" refers to in this context can be discerned through the haze of its language ("controlling," "eradicating," assessing "capacities," etc.), when we reach for a concrete meaning, it eludes us. It serves as a placeholder for whatever "management" will be at any moment. We see a handful of other plastic words ("plan," "information," "health," etc.) contributing to a vague tactic we are supposed to recognize as "invasive species management."

We need only investigate other disciplines to be made aware that "management" is a master of disguise with diverse personas. In the realm of economics, for example, "management" will rarely, if ever, refer to eradication. We can see very different implications for the word in the 2017 "Federal Trade Commission Draft Strategic Plan":

> Major Management Priorities and Objectives: The FTC's management objectives are incorporated into Strategic Goal 3: Advance the FTC's performance through excellence in managing resources, human capital, and information technology. This Strategic Plan addresses priorities in areas of human capital management, information technology management and planning, financial and acquisition management, staff emergency preparedness, records management and ethics.[49]

Here, the use of "management" generally refers to stimulation, advancement, and encouraged growth, rather than minimization or eradication. In these examples, the two meanings of "management" contradict each other, deflating each antonymic meaning. What might we now conclude on for a working definition of "management?" More money; fewer species. Capitalism in a nutshell.

Language is world-weaving, world-making. This is an old understanding of the weight of a word. Mythology, folklore, and origin stories across cultures tell of how speaking words can birth a whole world and its occupants. Words paint our world with color, culture, history, and context. It is through language and through words that people understand and interpret their worlds.

As we recognize the world-weaving power of language, we would do well to expose those places where institutions with power shrewdly gatekeep with words, wielding language to be selectively inaccessible. In these places, we must pay close attention and ask: Why is this so? How does this continue to be? And what can be done?

I do not mean to suggest that species introduced to new habitats that then cause great environmental damage are not harmful or that they do not contribute to biodiversity loss. I also am not arguing that humans have no place in shaping ecosystems or in trying to prevent biodiversity loss. In fact, I believe humans have a unique and important role as compassionate stewards to the land and its inhabitants—as evidenced in countless cultural traditions, particularly before and outside of capitalism. But these are social and ethical questions of environmental relationship that can't be left only to expert scientists and technocrats. These are conversations for society as a whole to learn about and engage in.

To this end, I explore how the prevalent story of invasive species (and their "management") is being told and the ways this story relies on misleading but persuasive rhetoric and surreptitiously inaccessible language. It is a story that undermines efforts to teach about and protect against the very harm (biodiversity loss, environmental degradation) caused by invasive species, which they purport to contest. I am not saying, "Let's not protect endangered species." Rather, I'm arguing that plastic words and militaristic metaphors are being used to confuse and make a spectacle of these various environmental issues while obfuscating the agendas of powerful institutions. In many cases across disciplines and throughout history, ways of thinking that are discordant with conventional opinion have little traction. This is equally true within the discourse on invasive species, and this book explores why.

The planet is undergoing a dramatic shift as part of an anthropogenic climate crisis. Many of our former paradigms are being thrown into question as climate crisis alters our world. This offers us a vital opportunity to change these sedentary ways of thinking—particularly those that have contributed to the climate crisis in the first place. We can assess prominent yet outdated narratives in favor of more holistic and thoughtful analyses. These paradigm shifts offer us a chance to build a more collaborative science.

Much as a statistical analysis of a small portion allows generalizations to be formed about a population, this book is similarly metonymic in nature. My in-depth analysis of this arguably niche debate around the language and rhetoric of invasive species management speaks to something much larger. It's a case study for recognizing science as an institution of power that influences how people think—how we think about ourselves, each other, and the world around us.

This book argues that language should be used accessibly, thoughtfully, and very carefully. I argue that certain disciplines of science must reevaluate their relationships with language and rhetoric, and that science must be made humbler to allow space for alternative solutions and ideas. At this point, we must ask ourselves, how did Western science come to have such a conflicted relationship with language? Let's turn now to the past to trace the history of this relationship and of how this scientific language was born.

CHAPTER 2

Science's Complicated Courtship with Language

Knowledge is, by its nature, always incomplete. "A scientist is never certain," the scientist Richard Feynman reminds us. And neither, the poet John Keats would argue, is a poet.
—Eula Biss

Science is engaged in the pursuit of truth. This is fundamentally different from saying science can always *tell* the truth.

Many scientific theories would be better understood as constructs that bear the imprint of the culture in which they were made. Science itself is a culture, one among many. The *Oxford English Dictionary* defines "culture" as "the cultivation or development of the mind, faculties, manners, etc.; improvement by education and training," and also "the devoting of attention to or the study of a subject or pursuit," and even "the distinctive ideas, customs, social behaviour, products, or way of life of a particular nation, society, people, or period," as well as "the philosophy, practices, and attitudes of an institution, business, or other organization."[1] The practice and study of the sciences fits neatly within these definitions. Cultures shift and morph; they are neither comprehensive nor universal.

Scientific findings—and the policies based on these findings—are often taken as reflecting natural law. It is true that science has revealed certain objective laws of nature (i.e., the law of the conservation of energy, the point at which water boils, the rate of gravity, etc.).[2] However, plenty of science-backed policies have historically been used to perpetuate sexism, racism, and oppression, with interventions that reflect less the laws of nature than the cultural biases of a time period. Phrenology, the study of human skull shapes, was once used to assert that Caucasian males were the intellectually superior manifestation of humankind—above all other races and genders. This purported to justify the claim that nonwhite people were biologically incapable of adapting to modern industrial societies. These findings conveniently coincided with US president Andrew Jackson's policy of Indian removal in 1830, which legally permitted the European colonizers of the Americas to forcibly and violently relocate local Indigenous peoples from their traditional territories.[3]

Similarly, "scientific" understandings of gender and sex have mirrored prejudiced beliefs of the time. We see this phenomenon unfold often today, as we are regularly besieged by transphobes and TERFs (trans-exclusionary radical feminists) with narrow, bigoted interpretations of chromosomal or other scientific markers of sex. Feminist scientist Donna Haraway remarks, "Women know very well that knowledge from the natural sciences has been used in the interests of our domination and not our liberation."[4] In Charles Darwin's *Descent of Man and Selection in Relation to Sex*, the idea that women are evolutionarily inferior is seminal to his text. "Man has ultimately become superior to woman," he argues, and not just in physicality and strength.[5] "Man is more courageous, pugnacious, and energetic than woman, and has a more inventive genius."[6] Similarly, in her article "The Scientific Revolution and The Death of Nature," environmentalist Carolyn Merchant describes how Francis Bacon—often referred to as the first philosopher of modern science and the father of empiricism—used sexually violent language to describe scientific study. Bacon spoke of nature as a woman keeping secrets "laid up in the womb" or hidden in her bosom, and he wrote, "A man [ought not] make scruple of entering and penetrating into these holes and corners, when the inquisition of truth is his whole object."[7] Bacon even encouraged his followers to grab nature by her hair and enslave her.

Merchant explains how Bacon went further and compared this violent way of extracting knowledge from nature to the way in which women suspected of witchcraft were subjected to torture in order to force a confession. Merchant argues that the interrogation and torture of witches was an early example of the scientific pursuit of knowledge. James I of England, she writes, "believed that witches had powers over people and nature, knew secrets, and could be forced to confess those secrets if interrogated under torture or shown the instruments of torture."[8] The scientific revolution of the sixteenth and seventeenth centuries, Merchant explains, was inspired in part by the desire for science, technology, and capitalism to "master" nature—by any means necessary.[9]

The wisdom of folk healers, midwives, and nurses—"witches" of the sixteenth century—did indeed pose a threat, but this was no threat to livestock, innocent babies, or pious Christians, as the puritanical myth would have us believe. Rather, their wisdom was a threat to the budding trade of male physicians, whose knowledge of health and medicine was sorely lacking compared to that of folk healers. Historian Clifford D. Conner, in his book *A People's History of Science*, quotes Bacon remarking that "witches and old women and imposters have had a competition with physicians."[10] It is therefore not surprising that the patriarchy of science sought the torturous eradication of these women and the violent extraction of their wisdom.

As environmental activist Vandana Shiva writes, "Western culture's favourite beliefs mirror in sometimes clear and sometimes distorting ways not the world as it is or as we might want it to be, but the social projects of their historically identifiable creators."[11] In short, there is plenty of evidence to show how certain interpretations of scientific "knowledge" reinforce harmful ideas of racism, sexism, and violence rather than unearthing an external, "objective" truth. We can add the literature of invasive species management to this list of misleading popular and scientific narratives. The idea that certain species are "invasive" emerged in the United States during a postwar period, as is evident in its persuasive war-heavy rhetoric that is still used today.

The words and beliefs of scientists, or their beliefs as interpreted by science journalists, carry a powerful influence over the public. In many ways, science strives to tell us who we are, why we are here, and what might happen next. But because scientific studies or findings can be impenetrable and even literally inaccessible to a general public, science also conceals much from us. This dual phenomenon of the influence and inaccessibility of science dates to the beginnings of institutionalized science and its complicated courtship with language itself.

History of Scientific Language

The seventeenth-century Enlightenment was fueled in part by the burgeoning of certain scientific societies, including the Royal Society in England, Rome's Academia dei Lincei, Germany's Academia Naturae Curiosorum, and France's Academie des Sciences.[12] In those days, scientists met in person to discuss their findings, but as science grew as a discipline, scientists began to write their findings in published papers. These read like modern-day tabloids—chronicling miracles like a sky with three suns or rains of blood.[13] Within two hundred years of the Royal Society's establishment, however, scientists had created a specialized technical language to report their findings. The result was a shared language understood by scientists of the day yet incomprehensible to anyone else. This lent science an aura of mystery: scientists were an elite, and their discoveries arcane and inaccessible. In this way, early scientific jargon was born, creating a dissonance between scientist and layperson that was often intentional.

Francis Bacon's short work of fiction *New Atlantis* evidences this idea that science should be reserved for a select group and its findings kept confidential. *New Atlantis* depicts Bensalem, a utopian island of vast technological achievements where a secretive school named Salomon's House enables an esoteric elite to hone their control and mimicry of nature. To the public, the purpose of Salomon's House "is the knowledge of causes, and secret motions of things; and the enlarging of the bounds of human empire, to the effecting of all things possible," but their findings and abilities are well guarded, known only to a few.[14] A note to the reader from Bacon's secretary explains that the author hoped Salomon's House would offer a practical model for existing and future schools for scientific knowledge in the late seventeenth century.

Phrenology, meanwhile, provides a nonfictional example of how exclusivity (in this case compounded by racist ideology) contributed

to the culture of science in the nineteenth century. The key phreno-
logical text *Crania Americana*, written by physician Samuel George
Morton, rationalized racialized hierarchies. Copies were only available to the richest institutions, but the book was widely read among
the educated class in the mid-1800s. Most people, however, were
exposed only to the racist violence of this ideology.[15]

Nevertheless, there were those who argued against the inaccessibility of the discipline and fought for scientific knowledge to
be disseminated democratically. During the scientific revolution of
the mid-seventeenth century, various groups opposed the monop-
olization of scientific knowledge in favor of "ending the distinction
between specialists and laymen" and forming a "new, democratic
system."[16] Gerard Winstanley, leader of a group of agrarian socialists
in England called the Diggers who rejected private property, "wanted
science, philosophy and politics to be taught in every parish by an
elected non-specialist," from the conviction that "knowledge and
education should not be monopolized by scholars in elite universi-
ties." But the language used to describe science remained technical
and the dissemination of knowledge undemocratic, while a conser-
vative Baconian science prospered.[17]

Contemporary examples of this foundational inaccessibility of
science are visible in the academic journals that are only accessible to
students or academics who are granted (or pay for) membership and
subscription fees that are prohibitive to most. The siloing of information makes it difficult to synthesize the research that already exists,
resulting in the creation of redundant studies. This also means that
many studies are not replicable, which goes against a fundamental
tenet of the scientific method.[18] While a certain precision and tech-
nicality of scientific language may be inevitable, a censored, guarded
dissemination of information is clearly not, as seen in the movement
toward open-access scholarship not sequestered behind paywalls.

A Dead Language

Latin has been the traditional language of science. Botanist Carl Linnaeus is known for carving up the world he observed using binomial
nomenclature and his taxonomic classification structure to define

organisms, which gave animals and plants specific Latin names. Latin was once spoken in Latium (modern-day Rome), but no existing cultures on earth use it for vernacular communication, making Latin a dead language.

At first glance, this use of Latin appears to be a way to write about science objectively. There was no one culture that could claim Latin as its own tongue, and as Latin belonged to no one, it could belong to everyone equally. Instead, it was only the colleges and schools that taught Latin, meaning that the uneducated masses would not be able to understand anything written in this dead language. There is also historical evidence that suggests that Latin was intentionally chosen to elevate its subjects: historian George Sarton notes that Latin "was the esoteric language used to prevent the dissemination of learning to people who were deemed unworthy of it, or who might make a bad use of it."[19] Italian natural philosopher Giambattista Della Porta, an early experimental scientist in the 1560s who imbued folklore and magic in his scientific pursuits, primarily "wrote in Latin, and not for the people."[20] Around this same time, the Protestant Reformation was unfolding, during which German monk Martin Luther challenged ordinances of the Roman Catholic Church.[21] One lasting change was that mass would no longer be spoken only in Latin. Traditionally, priests would convey sermons entirely in Latin, which meant that most people could not understand them. In an attempt to make mass more accessible to churchgoers, sermons would be spoken in the vernacular instead.[22]

Although Francis Bacon may have insisted that the study of science was to better mankind, he is quoted as saying, "I do not like the word People," who he referred to as "the commonality" and "the meaner sort."[23] And while Bacon and other scientific revolutionaries of this time would occasionally write in the vernacular as well as in Latin, their publications were designed for the educated and the urban middle class, who together would form the dominant elite over an uneducated working class.

Distrust of Language

Ultimately, scientists grew to generally distrust language itself. In their essay "The Image of Objectivity," science historians Lorraine Daston and Peter Galison quote a late-nineteenth-century physiologist explaining, "Born before science, language is often inappropriate to express exact measures or definite relations."[24] Or, as Roland Barthes explains in his essay "Science Versus Literature": "As far as science is concerned language is simply an instrument, which it profits it to make as transparent and neutral as possible; it is subordinate to the matter of science (workings, hypotheses, results) which, so it is said, exists outside language and precedes it. On the one hand and first there is the content of the scientific message, which is everything, on the other hand and next, the verbal form responsible for expressing that content, which is nothing."[25]

An indeterminate phenomenon, as the theory of semiotics has shown, language was thought unable to reflect the precision that science's accurate revelation of nature seemed to require. Language can be laden with multiple meanings and thus open to interpretation, at odds with the exactitude of science. In the late nineteenth century, scientists cultivated the concept of "objectivity" in order to make science more precise. This objectivity was anchored in the scientific method, a series of principles for the systematic approach to acquiring knowledge. According to the scientific method, one needed to create a hypothesis based on an observation, test it through experimentation, and then analyze the resulting data to form conclusions and provide evidence using things like charts and graphs. The new technology of photography, invented in the 1820s, was especially well suited to render nature precisely, without the messy mediation of language. Objectivity in science, as Daston and Galison explain, was considered a scientist's moral duty. They needed to do their utmost to identify and set aside their biases, expectations, and emotions in favor of a more mechanical way of representing nature. Through more objective methods of scientific inquiry, it was believed that nature might be better able to speak for itself. Language betrayed humanity, and so scientists trusted photography and related technologies not to tell but instead to show nature's truth.[26] The assumption was that subjectivity should not affect scientific findings or agendas, and this assumption largely remains in place today.

Argentinian writer Jorge Luis Borges satirically references this desire for precision and accuracy in his one-paragraph short story "On Exactitude in Science":

> In that Empire, the Art of Cartography attained such Perfection that the map of a single Province occupied the entirety of a City, and the Map of the Empire, the entirety of a Province. In time, those Unconscionable Maps no longer satisfied, and the Cartographers Guilds struck a Map of the Empire whose size was that of the Empire, and which coincided point for point with it. The following Generations, who were not so fond of the Study of Cartography as their Forebears had been, saw that the vast Map was Useless, and not without some Pitilessness was it, that they delivered it up to the Inclemencies of Sun and Winters. In the Deserts of the West, still today, there are Tattered Ruins of that Map, inhabited by Animals and Beggars; in all the Land there is no other Relic of the Disciplines of Geography.[27]

Borges hyperbolically describes what happens when the map represents that which it depicts on a one-to-one scale: the map becomes the territory itself and is both cumbersome and useless. This playful thought experiment satirizes the impracticability—to the point of nonsense—of an exact correspondence between signified and signifier: there is always slippage. German researcher and writer Wolfgang Sachs finds a modern equivalent of Borges's cartographers in a description of what satellite images can and cannot relay:

> Satellite pictures scanning the globe's vegetative cover, computer graphs running interacting curves through time, threshold levels held up as worldwide norms are the language of global ecology. It constructs a reality that contains mountains of data, but no people. The data do not explain why Tuaregs are driven to exhaust their water-holes, or what makes Germans so obsessed with high speed on freeways; they do not point out who owns the timber shipped from the Amazon or which industry flourishes because of a polluted

Mediterranean Sea; and they are mute about the significance of forest trees for Indian tribals or what water means in an Arab country. In short, they provide a knowledge that is faceless and placeless, an abstraction that carries a considerable cost: it consigns the realities of culture, power and virtue to oblivion. It offers data, but no context; it shows diagrams, but no actors; it gives calculations, but no notions of morality; it seeks stability, but disregards beauty.[28]

Borges's cartographers' mapping technique provides a hyperaccurate representation of the landscape but is only two-dimensional, lacking in depth. In their attempt to be perfectly objective, the cartographers, as well as our satellites, fail to be *empirical*—erasing and occluding facts that don't fit their measurements.

Similarly, science seeks to function through an abstraction from the world and as such misses much. This, regrettably, is the ultimate goal in the pursuit of objectivity. Not only is objectivity essentially impossible, the "objective" is always provisional, a model that can show only a partial perspective. In the nineteenth century, the quest for objectivity resulted in the valorization of detailed, "scientific" drawings as well as photography. There was Ernst Haeckel, whose famous zoologic illustrations depict colorful mosaics of sea creatures, mammals, and botanicals. John James Audubon came to fame for his detailed paintings of birds. Anna Atkins captured ghostly white shadows of her botanical collections using cyanotype. Before the concept of scientific objectivity was born, there was William Hunter, who drew extremely precise and detailed anatomical portraits of the human body. According to Daston and Galison, Hunter did not "regard the aesthetic with suspicion, as being opposed to scientific accuracy" as later science objectivists might. Rather, he "considered the beauty of depiction to be part and parcel of achieving that accuracy, not a seduction to betray it."[29]

What Daston and Galison show is that these techniques of portraying nature are not objective, they are truly stylistic or aesthetic preferences. In his essay "Newton and the Ideal of a Transparent Scientific Language," author J. M. Coetzee argues that science's preference for the passive voice is similarly a stylistic mode by which subjective views come across as objective, given, natural, and eternal. Rather than "I observed a certain natural phenomenon," which implicates a human observer, we have instead "A certain natural phenomenon was observed," which suggests that anyone, anything, even an omniscient god, would have observed the exact same natural phenomenon in precisely the same way.

But this is not quite possible. Philosopher Immanuel Kant explored how bias inherently cloaks our understanding of the world. According to Kant, our minds organize the world around us from mystery and chaos into something orderly and familiar. This organization differs from person to person, mind to mind. Therefore, the true world is likely not as we see it or understand it, because each of us sees the world differently from everyone else. Ultimately, the objective world exists beyond our inherently subjective minds.[30] This is not to say that there is nothing that science can tell us about our universe; rather, many of the unfolding stories that science tells us must be up for debate or taken with a grain of salty nuance.

The Effect of Objectivity on Language

What are the effects of a scientific technical language born out of this desire for objectivity? One effect is diffusion of science's inhuman "objectivity" into language itself. Richard Stivers, in his book *Technology as Magic*, describes how science builds its vocabulary. Lexicographers, he explains, have determined that science and technology are responsible for nearly half of the new words added to the English language in the twentieth century.[31] Uwe Poerksen argues that "a scientist is fundamentally the master of his language." The scientist might use symbols, phrases, or abbreviations aiming to convey something "briefly and unambiguously," but these may not be applicable to other usages.[32] The scientist therefore enjoys a language made for exactly this purpose, a language that is not *common*. Though current scientific vocabulary may no longer draw just from Latin and is more widely shared with the "meaner sort" in a way that would greatly disappoint Bacon, the obscuring aspect or function remains: science and the common vernacular are still largely kept separate.

In *The Human Condition*, German philosopher Hannah Arendt addresses what it means for science and language to remain distinct, for logic and reason to leave behind the poetic, human, familiar, and relatable. She notes that "the 'truths' of the modern scientific worldview" can be "demonstrated in mathematical formulas and proved technologically;" that "a 'language' of mathematical symbols" originally meant as abbreviations of or placeholders for common language can now be strung together to form "sentences" untranslatable into any vernacular. Arendt expresses concern about the political implications when ordinary people are unable to talk or think about what they are doing, instead requiring "artificial machines to do our thinking and speaking." Because we are so often prompted to adopt the ways and rites of science into our daily lives, Arendt believes that in the end "we would in all earnest adopt a way of life in which speech is no longer meaningful."[33] Speech and the rules of grammar become less important in the pursuit of the rules of nature, and clarity is therefore sacrificed.

The Effects of Scientific Jargon

There is a name for the mechanical language Arendt and Poerksen describe, the language scientists create when they invent new concepts and need to give those concepts a name that lies beyond common day-to-day language: jargon.

In the *Oxford English Dictionary*, "jargon" is defined as "unintelligible or meaningless talk or writing; nonsense, gibberish. (Often a term of contempt for something the speaker does not understand)," "a conventional method of writing or conversing by means of symbols otherwise meaningless," and "applied contemptuously to any mode of speech abounding in unfamiliar terms, or peculiar to a particular set of persons, as the language of scholars or philosophers, the terminology of a science or art, or the cant of a class, sect, trade, or profession."[34]

Specialized terminology suggests professionalism. When scientists discover new findings, they sometimes create terms to describe these findings. These new, highly specific words will not be used by everyone—they will almost entirely be used only by those working in the specific scientific discipline in which the new word was born. In principle, jargon helps condense otherwise wordy explanations.

In specialized disciplines, jargon is effective in communicating new findings. Acronyms can be used for shorthand, and fewer words can be used to describe more complex phenomena. But in exchange for brevity, clarity is sacrificed. For example, with scientific jargon, one could describe cellular activity in this way: "The phospholipid bilayer allows for bidirectional transport of cellular metabolites via membrane pores and transmembrane proteins." Alternatively, the same activity could be explained this way: "The cell membrane allows for the entry of molecules needed by the cell as well as the exit of molecules produced by the cell. Depending on the molecule, it will either pass through small holes in the membrane called pores or through proteins embedded in the membrane."[35] The difference in comprehension between the two sentences is clear. Anyone who is not a specialist in cell membrane activity will be able to understand the longer second sentence with more ease than the first.

For the general public, this jargon confuses. It distracts. It complicates and frustrates. This compartmentalization of language says, "Leave it to the experts." Similar to plastic words, jargon makes the discourse of science largely inaccessible and therefore unassailable, as readers struggle to grasp concepts when they are so abstracted from their original context.

It is estimated that in general academic texts, there is 5 percent jargon, 80 percent high-frequency or commonly used familiar words, and 8 to 10 percent academic vocabulary. *Scientific* academic texts, however, contain around 22 percent jargon. Meanwhile, according to a study by Marcella Hu and Paul Nation, accurate comprehension of a document requires 98 percent familiarity with the vocabulary—impossible if only 78 percent of the words of a scientific text are clear to your average reader.[36] When the Flesch Reading Ease (FRE) test—which measures a text's readability from 0 (unreadable) to 100 (understandable)—was applied to "Summaries for Policymakers" from the Intergovernmental Panel on Climate Change, the document scored below 20, a dishearteningly low score for an organization that is chiefly tasked with monitoring research in global climate crisis and effectively sharing that information with the general public.[37]

These concepts, rendered in unfamiliar and idiosyncratic grammar and syntax, become, in effect, a foreign language—unreadable and unspeakable by the general public, for whom the scientific subject matter nevertheless has impacts. This creates a form of illiteracy that forecloses politically consequential public engagement with scientific problems, defending science against comprehension, contestation, or resistance.

Creating new words is fundamental to keeping language alive and relevant as our values, beliefs, and ways of communicating with one another change around us. However, scientific jargon can be dehumanizing in its abstraction. It diminishes the scientist's ability to communicate appropriately and effectively with other people. There are times when this gap in understanding can have distressing consequences. For example, when a genetic counselor meets with expectant parents to discuss the potential outcomes facing their unborn child, there is a dissonance between the concepts that need to be relayed to the pregnant person and their own understanding of their baby. Geneticist and philosopher Silja Samerski describes this effect in the context of genetic counseling. She writes: "On the one hand, [the word] 'gene' demands significance, meaning. On the other hand, it cannot signify anything outside the lab. This paradox enables the 'gene' to work as a bridge between statistics and real life."[38] Samerski goes on to relay a scene in which a genetic counselor seeks to advise a pregnant woman:

1. The geneticist talks to a laywoman. He has to spell out his knowledge in such a way that normal people can follow him. To do so, he has to find everyday words for notions like chromosomal aberration, DNA-mutation and probability model.

2. Once talked to, the client is urged to make a decision. This decision is, in some way, a decision about life and death, about delivering a child or terminating a pregnancy. Facing the counselor's genetic mumbo jumbo the client inevitably asks herself: What does all this say about me? What does all this mean to me? Genetic counseling is a glaring example of the clash between scientific concepts and everyday meaning.[39]

This is an example of a delicate situation in which both parties would greatly benefit from sensitivity to and skillfulness in translating between specialized jargon and the ordinary vernaculars of laypeople. We can see here how the "mumbo jumbo" of jargon stokes confusion, dissonance, and distance between the pregnant person and the genetic counselor.

Objective abstraction from human emotion and bias in this sort of situation is emotionally devastating. To the would-be mother, her could-be child has been transformed before her eyes into a frightening, dangerous risk. Dehumanizing, indeed, and also disempowering. The mother is rendered powerless at the hands of the genetic counselor. Her knowledge of her own body, womb, and baby is inconsequential next to the knowledge of the scientific expert, the "genetic counselor."

Scientific jargon isn't itself necessarily problematic. In fact, jargon is unavoidable. However, the abstract, inhuman way of relaying information as scientific jargon can be startling and manipulative. We can see here how the term "gene" has the power alienate, to cause confusion and pain.

In many cases, like the example above, doctors and scientists learn the technical language of their discipline without equally valuing what is called rhetoric, the idea that *language itself* is a social act.[40] There is another bias that emerges when relaying technical or scientific information, a bias called the "curse of knowledge," which describes the tendency for people to forget they once did not have the knowledge they do now.[41] This "curse" makes it difficult for them to recognize the distance between themselves and a less-informed audience. It serves as an obstacle for scientists to empathize with their audience or to accurately consider clarity in communicating what their audience may not understand.

Plastic Words and Scientific Jargon

Unlike the vernacular origins of plastic words, scientific jargon emerges from scientific domains and rarely mingles with vernacular language. Plastic words are general and vague, while scientific jargon is highly technical and specific. Terms in scientific jargon retain their meanings consistently in their context, while plastic words are malleable and morphable. However, while they appear dichotomous in these respects, their exclusionary effects are similar in that they abstract and distance. Plastic words and scientific jargon are both hard to translate broadly. However, when we hear scientific jargon, we do not presume to understand it (unless we already count ourselves among the experts in the discipline). Compared to scientific jargon, plastic words are a much trickier lexicon, largely because of their widespread use. We tend to think we know what plastic words mean across any number of contexts—but they hide in plain sight.

The technical tongue of jargon—whose history is steeped in elitist remove and intentional obfuscation—and these imprecise, interchangeable plastic words make scientific findings difficult for a general audience to understand.

Today we live in what is often called the information age, which suggests that we have easy access to information. True, "information" can be accessed with near instantaneous speed almost anywhere in the world, which in turn holds the potential to let scientific findings reach many more people. But while the public can easily access scientific findings, they may not have equal access to *understanding* what these findings mean.[42] I argue that scientific texts must urgently emphasize clarity and readability, and that it is possible to do so without compromising professionalism or abandoning necessary terms.

Our Culture and Science Communication Today

Our dominant culture valorizes science. Many of us see science as something that makes our lives easier, tells us how to live longer, and unveils hidden mysteries of the universe. At the center of this powerful discipline is the fact that most of us cannot speak its language.

Some may argue that it is the job of popular science journalists, not scientists, to translate scientific findings to the public. Writers such as Malcolm Gladwell, Atul Gawande, Michael Pollan, and many others take complex scientific issues and rearrange them into manageable, simplified, energized forms for nonscientists. The problem with this popular science writing is that it is not always translation of information; it is often oversimplification. The effect is leading a mass audience to believe they understand a topic that they may actually not understand.

In March 2018, the *New Yorker* published an article by Kathryn Schulz describing a well-known invasive species: the stink bug. The article vividly paints these insects as swarming entities straight out of a horror film. Its grotesque imagery is captivating, and many letters to the editor praised the author's style. "I simply could not tear myself away," one wrote. "What a disgusting story, brilliantly written: Stephen King meets Rachel Carson," wrote another.[43] Yet one reader balked:

> As Mark Twain noted, "Nature knows no indecencies; man invents them." Schulz relies on the same highly militarized and villainizing language that's widely used to describe other so-called invasive species. These critters have arrived in their new homes not of their own agency but through careless (and sometimes intentional) handling by humans. Our breathtaking sense of exceptionalism ensures that our errant ways in bringing pests from elsewhere is never our fault but somehow that of the organisms themselves. They are blamed for doing what all organisms do—attempt to reproduce and survive. As loathsome as they might smell, act, or be, they are not the villains in these environmental dislocations; we are.[44]

When dealing with science, which has no small amount of influence over policy or politics, we must ask if simplification is a service or manipulation. In this case, I argue the metaphor does a disservice—it neglects to unpack deeper nuances of the invasive species story that at least one careful reader has begun to dig into.

While most, if not all, scientific studies suffer from some degree of inaccessibility, the attempts to convey the disasters of climate crisis are notably met with pushback—not because these studies are rhetorically misleading but because of an ideologically driven mass denial. In his terms in office, US president Donald J. Trump used much of his power to roll back regulations designed to curb emissions and other aggravators of climate crisis. He also fostered a cult following of uncritical science deniers who were then largely responsible for the anti-vax movement during the early years of the COVID-19 pandemic.[45] This widespread mistrust of science and of the educated elite seems to me to be a misplaced rebellion against science's own historical mistrust of the uneducated masses, and a symptom of elitist communication in and around science today.

Unfortunately, is not uncommon for some to fudge results—or at least manipulate how they are presented to fit a preferred narrative. Scientists, researchers, and pop-science writers alike have been known to share misleading findings, employ p-value hacks, ignore peer-reviewing and replication standards, or discard results that don't offer the conclusion they are hoping to reach. Consider the research conducted on the efficacy of antidepressant medications in 2022. Many antidepressant manufacturers have funded research into their own drugs themselves, which would seem an important conflict of interest. After many trials, only the trials that suggest significant efficacy are kept; the rest (which are often the majority) are trashed—this is known as reporting bias.[46] "Using FDA reviews on 4 newer antidepressants, we identified 30 trials, half with positive, and half with negative, outcomes. Among the 15 negative trials, 6 were unpublished and 2 others were misreported as positive. Seven other negative trials (47%) were reported transparently (as negative), an improvement over the low (11%) rate found earlier with the older antidepressants." This suggests that while reporting bias "appears to have diminished for newer, compared to older, antidepressants," it nevertheless "remains a significant impediment to researchers and medical decision-makers."[47] Writer and journalist Johann Hari discusses this phenomenon in his book *Lost Connections: Uncovering the Real Causes of Depression—and the Unexpected Solutions*. He writes, quoting physician and scientist John Ioannidis,

> "The companies are often running their own trials on their own products," he said. That means they set up the clinical trial, and they get to decide who gets to see any results. So "they are judging their own products. They're involving all these poor researchers who have no other source of funding

... [and who] have little control over ... how the [results] will be written up and presented." Once the scientific evidence is gathered, it's not even the scientists who write it up much of the time. "Typically, it's the company people who write up the [published scientific] reports."[48]

Obviously, profit is a compelling motivator.

The falsification of data occurs across scientific fields. A 2009 study suggests an average of 2 percent of scientists admitted to fudging data at least once in their career. A more recent study suggests that one in seven scientific papers (about 14 percent) may be at least somewhat fabricated. These studies both ought to be taken with a grain of salt, suggests James Heathers, psychology researcher and orchestrator of the newer study. Their methodology relies "on asking researchers directly if they had engaged in dishonest research practices, Heathers said, 'which I think is a very bad approach to being able to do this.' But he noted that it was probably the only method available to use when the research was conducted. 'I think it's pretty naive to ask people who are faking research whether or not they'll honestly answer the question that they were dishonest [about] previously,' Heathers said."[49] A 2018 CBC News article discusses a similar study by epidemiologist Ralph Katz in which Katz attempted to figure out how frequently statisticians falsified data. "'Greater than 20 per cent, sometimes up to 50 per cent, said these had occurred a multiple number of times over the last five years,' said Katz, adding that this does not mean all research is compromised. 'There are a lot of consultations being done where this is not happening, but it is alarming how often it is happening.'"[50] Of course, there are plenty of well-conducted experiments and researchers and institutions that share fundamentally important scientific findings with the public. I am not a science denier, and I hope these examples do not feed that problematic narrative. At the same time, the chilling trend among some scientists to intentionally offer misleading (if not outright false) results decreases public trust and creates a breeding ground for increased science denial. I offer this critique with the hopes that the institution of science will improve its credibility and prioritize public comprehension.

It is worth noticing that Trump employs the same inaccessible plastic words in his denial of science that the scientist uses to convey their findings. Trump speaks with a forked tongue to promote his own alternate narrative against a scientific agenda. In his 2017 speech withdrawing the United States from the Paris Climate Accord, he argued that the accord diminishes *production* and reduces *development*.[51] As Uwe Poerksen writes, "These opaque vocabularies above all shout 'science!'"[52] The words Trump uses are ironically cloaked in the very same sort of scientism he argues against. Because plastic words do not carry any meaning, they can be used equally by both proponents and deniers of science to articulate their claims. Plastic words ultimately do a disservice to science and make meaningless blather like Trump's (exemplified here) sound important, accurate, and correct. They in fact undermine science. With these words, even the weakest argument can gain momentum and power.

These episodes evidence a strange parallel between how people engage with science and how people engage with religion. With religion, there will be devout evangelists and selective believers as well as nonbelievers or believers in other religions. Similarly, in science, we have uncritical followers, selective followers, and outright science deniers (a group that has a large overlap with devout religious evangelists). But science is not religion. It is not meant to profess the absolute truth, only to pursue it. Science as a mode of systemized critical inquiry should be the opposite of dogma. So what is it about the way science manifests in our society that leaves it open to the formation of dogmas?

I would suggest a fundamental rethinking of science's place in culture. I believe that to humble science is to elevate it. It is important not just to admit but also to emphasize that this belief in understanding nature and reality through science in fact rests on a degree of uncertainty.

.

CHAPTER 3

The Rhetorically Charged, Ethically Dubious, Xenophobic, and Capitalistically Motivated Story of Invasive Species Management

A recipe for Dandelions: A lungful of air will spread the seeds.
Manicured lawn becomes bed of weeds.
—CrimethInc.

The dominant story of invasive species in Western science tells us that they are fundamentally enemy invaders and must be battled and ideally eradicated from terrain to which they do not belong.[1] But there are many working in this field who argue against the inherent bias of that story. Critically, the history of invasive species is also the history of empire. It is possible to observe and study invasive species and their ecology with nuance and scientific rigor. There are scientists, researchers, writers, activists, and others pushing for new ways of thinking about these commonly misunderstood species.

In his book *Invasion Biology*, biologist Mark A. Davis contests this dominant narrative and historicizes how invasive species have come to be defined. The earliest recognition of an invasive species is undoubtedly impossible to determine. Species migrate and spread and have done so for millennia. When humankind generally ceased its nomadic habits and settled, it is not unlikely that, through observation of their surroundings, our ancestors noticed the effects of introduced species established through migration or trade.[2] However, the first known reference to the concept of nonnative species in Western science occurred in the eighteenth century during the end of the age of discovery, a time defined by European expansion and colonization. The Swedish botanist and naturalist Peter Kalm, a student of Carl Linnaeus, hoped to bring nonnative species to Sweden to grow for profit.[3] In the course of his explorations, he recorded the presence of European flora and fauna in North America. Likewise, geographer and naturalist Alexander Humboldt recorded a global spread of certain species during his travels.[4] Early encounters with these migrated organisms were referred to with an air of curiosity, exciting discovery, and a desire to document and remember.

The Beginnings of the Militaristic Metaphor in Invasive Species

The emergence in the eighteenth century of biogeography—a discipline that focuses on the global migration of flora and fauna—turned what had previously been travelogue and anecdote into a more systematic examination of native and nonnative species. Biogeography divided the globe into six distinct regions, each characterized by differences in climate and environment. This approach may have helped bring about the concept of permanent *native* species—a complex claim given the tendency of species to regularly and obviously migrate between regions and across borders.[5] Yet, in this new order, species were now decreed to have originated from a singular location.

Botanist John Henslow is thought to have coined the term "native species" in 1835, and its use rapidly grew. Just over a decade later, botanists began using the terms "alien" and "native" to describe plants.[6] In the 1860s, there was recognition and discussion among agricultural scientists in the United States that nonnative species could be harmful to native species, but so too was there a desire to "procure, propagate and distribute . . . new and valuable seeds and plants" in the interest of agricultural production, according to the founding mission statement of US Department of Agriculture (USDA) in 1862. By the early 1900s, the harmful effects of particular nonnative species were more widely understood (specifically the introduction of pests that would harm agriculture) alongside a growing xenophobic mentality that distinguished between native and nonnative plants and showed "enthusiasms of the native, and fears of the alien." The militarization of the botanical lexicon began to emerge at this time, particularly with the initiation of the Plant Quarantine Act of 1912 and the establishment of the Plant Quarantine and Control Administration of 1928.[7] These agricultural preservation policies were the beginning of a singular obsession with fighting against invasion by nonnative species. Conservation of native species originated to protect against biodiversity loss and began to take on a militant overtone alongside the increasingly polarized narrative of native versus nonnative species over the course of the twentieth century and into the twenty-first.

This insistence on designating certain contained areas for species that emerged from a specific biogeography echoes the constraining of borders designed to keep people in their own countries and out of others. In *Borderlands/La Frontera*, feminist theorist Gloria Anzaldúa describes the othering effects of such borders: "The U.S.-Mexican border *es una herida abierta* where the Third World grates against the first and bleeds. And before the scab forms it hemorrhages again, the lifeblood of two worlds merging to form a third country—a border culture. Borders are set up to define the places that are safe and unsafe, to distinguish *us* from *them*."[8] This government-funded separation between lands emboldened an obsession with borders—which lie between *us* on one side and *them* on the other—and would perpetuate a xenophobic or nationalistic way of thinking, even in the realm of environmental science.

Xenophobic Effects

In *Confronting Introduced Species: A Form of Xenophobia?*, ecologist and evolutionary biologist Daniel Simberloff documents unsettling examples of nationalism and xenophobia in invasive species management. In 1992, German garden architects Joachim Wolschke-Bulmahn and Gert Groning documented "a Nazi campaign to cleanse the German landscape of unharmonious foreign substance [plant species]," with glaring ominous parallels to the party's ideology of ethnic cleansing.[9] Simberloff also recounts a Danish immigrant to the United States in the early twentieth century who, having initially embraced nonnative species "like Russian olive (*Elaeagnus angustifolia*) and tree of heaven (*Ailanthus altissima*), generally on grounds of hardiness," later grew to oppose them.[10] He longed for his gardens to exist "in harmony with their landscape environment and the racial characteristics of its inhabitants" and for them to "express the spirit of America and therefore . . . to be free of foreign character as far as possible." In speaking of certain plant species, he bemoaned how "the Latin and the Oriental crept . . . more and more over our land, coming from the South" and said that "the Germanic character of our race, of our cities and settlements was overgrown by foreign [character]." "Latin," he argued, "has spoiled a lot and still spoils things every day."[11]

Science historians Lorraine Daston and Peter Galison describe the infamous racial atlases created by Nazi Julius F. Lehmann, who published medical texts about genetics, eugenics, and "racial hygiene" and also created medical atlases for the journal *Munich Medical Weekly*. He later found success creating atlases that described the physical characteristics that could determine race. Lehmann wrote in 1920 that he wanted to publish a "human field guide to the flora (*Excursions flora*) of Germany that, first of all, would lay out the general racial markings in an exemplary fashion."[12] As in the study of phrenology, these racial atlases were inspired by the premise that, if plants belong to a particular hierarchized taxonomy, why wouldn't people? During World War II, the Nazis would rely on Lehmann's work in their own genocidal agendas.

German pest control company Deutsche Gesellschaft für Schädlingsbekämpfung mbH (German Society for Pest Control, shortened to Degesch) was founded in 1919 and produced insecticides and fumigants. In 1922, Degesch developed a variant of the cyanide-based pesticide Zyklon, which had been banned following its use in WWI. This variant, which the company patented as Zyklon B, was sold in 1942 to the SS, who used it to murder more than one million people in death camps. The company's managers would later be executed for war crimes committed during the Holocaust, after a British war tribunal determined they had knowingly supplied the chemical for this purpose. After World War II, the poison changed names, but it is still in use as a pest control agent. These examples display a chilling overlap between the environmentalist impulse (and even methods) for eradicating unwanted species and the nationalistic push toward xenophobic genocide.

After World War II, American scientist Charles Elton (who would become known as the father of invasive species management) published *The Ecology of Invasions by Animals and Plants*. Its first sentence—in its first chapter, titled "The Invaders"—exemplifies this militaristic mentality: "Nowadays we live in a very *explosive* world, and while we may not know where or when the next *outburst* will be, we might hope to find ways of stopping it or at any rate, *damping down its force*."[13] Published in the postwar era, Elton's book is charged through with visions of omnipresent war and enemies. Elton's text inaugurated a fearful urgency in discussions of invasive species, which subsequent texts paralleled. While postwar trauma fueled this militaristic mentality, perhaps there is also the echo of a settler-colonial preoccupation with the fear of being invaders themselves.

It took a quarter of a century after the book's publication before "invasion biology" began to emerge as a new discipline, alongside the establishment of the Scientific Committee on Problems of the Environment in 1983. Since the 1980s, however, interest in the field has surged.[14] The Google Books Ngram viewer shows an exponential increase in the use of the phrase "invasive species" starting in the mid-eighties and peaking around 2008. While it took time for Elton's book to gain popularity among scientists and the general public, the militarism of his vision for these sorts of species was undoubtedly foundational to invasive species management.

The Military Metaphor in Invasive Species Management

Ultimately, the militaristic metaphor is one of conformism. It reminds us of a deep (perhaps in some cases instinctual) fear of the other, of that which is unknown or different. We are afraid of what we don't know because we fear it might harm us, and the metaphorical militarization of invasive species feeds on this fear. Horror writer H. P. Lovecraft declared, "The oldest and strongest emotion of mankind is fear, and the oldest and strongest kind of fear is fear of the unknown."[15]

Militaristic language may serve to colloquialize concepts like invasive species and rally popular support against them. Yet this aggressive, combative approach enforces the false notion that human beings are somehow separate from or against certain manifestations of nature.[16] As environmentalist writer Wendell Berry says in his essay "Getting Along with Nature," "Even as conservationists, we see the human and the natural economics as necessarily opposite or opposed; we subscribe to the very opposition that threatens to destroy them both."[17] While our capitalist society enforces patterns or behaviors that distance us from the natural world, we are and have always been a part of nature.

Biologist Brendon Larson explores problems with these military associations in his article "The War of the Roses: Demilitarizing Invasion Biology," explaining how our current rhetoric relies on the power of metaphor to generate a unanimous consensus on the meaning and significance of invasive species. This can be misleading to the public, as it marks these species as malicious entities. Larson also notes how this more violent way of thinking diverges from the prevailing philosophy in other domains of environmentalism or ecology that strive for a more harmonious coexistence between humankind and nature as part of broader conservation values. In fact, the sense of urgency invoked for invasive species management procedures through militarized metaphor can undermine conservation efforts. Larson sees this as a multifaceted problem, and I paraphrase:

1. The metaphor of war requires one to imagine two opposing sides and to harbor an us-versus-them mentality. This is contradictory, given that invasive species are largely an endemic problem because of anthropogenic factors like human globalization. The *us* in this war facilitated the spread of *them*. We cannot be separated; we are the ones creating the invaders. The use of a war metaphor pits us as opposing sides, but our relationship with invasive species is complex and inherently entangled.

2. The metaphor of war suggests that there will be a winner, which is misleading. There will never be a return to pure natives in an ecosystem, particularly as climate crisis and other anthropogenic factors continue enabling dispersal.

3. Such loaded military language may contribute to public resistance to or distrust of those using this language. Science is supposed to be objective, and loaded language bears bias. A Pacific Northwest–based study conducted on the role of ecologists in ecological

management suggested that the public prefers ecologists to inform and educate rather than argue for a particular mode of action. 4. Invasive species removal in certain areas of conservation or restoration will primarily benefit upper-middle-class citizens and may therefore be perceived as an elitist ecological pursuit. This class divide further distances the public from ecologists and conservationists and perpetuates public mistrust.[18]

Invasive species management with a militaristic logic will fundamentally prioritize control and oppression rather than conservation. The dichotomy between humans and nature can, in a way, free us from accountability or connection to nature's struggles. With substantial debate around these topics kept at a linguistic and discursive remove and accessible only to experts, the public is rendered uninformed and unable to insist how our natural world, our collective home, might be taken care of.

Elsewhere, Larson argues that the current invasive species paradigm perpetuates an obsolete way of thinking about ecosystems as stable systems rather than as changing entities eternally subject to ecologically necessary disturbances. Larson pushes emphatically against the Cartesian dichotomy between humankind and nature, challenging us to recognize our role in the creation of scientific paradigms. He quotes Finnish environmental policy scholar Yrjö Haila: "Humans are creatures of nature; consequently, discriminating between phenomena of nature as 'natural' and phenomena of culture as 'unnatural' does not make sense at all."[19] At what point does human manipulation of nature become unnatural? There is no definitive answer to the question of how nature and culture relate; in fact, it is one of the most contentious questions in all of academia. "Like it or not," Larson writes, "this hybrid world is the one in which we live."[20] Larson likewise warns that the militarized discourse surrounding invasive species fosters a generic sense of otherness, with an invasive *other* linguistically made to seem as if it falls outside of our natural world.[21] He writes, "Invasion biology relies upon an underlying dichotomy between nature/non-human/ native and culture/human/nonnative that exemplifies the exclusion of human beings from ecological systems. Invasion biologists thus focus on species that *we* have introduced and which cause harm, rather than native ones that have similar effects. Their 'naturalness' thus becomes the focus of ecological debates. In a recent exchange about the significance of invasive species, for example, [one study] emphasized the distinctiveness of recent human-induced changes, whereas [another] decried those who claim they are unnatural."[22]

It is imperative to dismantle this false separation of us versus them, of us versus nature, of us versus invasive species. This belief leads us only to an unwinnable war, with losses on both sides.

Ecocidal Eradication Efforts

In late spring of 2020, enjoying freedom from work during the COVID-19 lockdown, my partner and I wandered around a park we love, called Grass Lake. It's got a big meadow and a thicket of forest surrounding a marshy lake. The meadow is usually scattered with rows of bushes of Scotch broom, considered invasive where we live. Every so often, parks department employees spray and bulldoze in an endeavor to eradicate them. On that late spring afternoon of the first plague year, we arrived after one such attempt and were aghast. The park looked terrible; it looked sick. It had been poisoned and trampled, after all. My partner squatted to get a closer look at the damage and found a dead baby garter snake—an innocent (and native) victim of these redundant and seemingly pointless, violent removal efforts. Wandering around in shock, I couldn't help but feel furious at how the whole ecosystem itself, including its native species, is routinely under attack all for the sake of what some call "invasive species management." The health of the entire ecosystem was considered worth sacrificing in these Sisyphean eradication efforts—and yet the Scotch broom always comes triumphantly back.

While Scotch broom is frowned on for its tendency to spread broadly and outcompete native plants, Pacific Northwest folk herbalist Corinne Boyer argues that it has benefits often ignored due to its invasive status. "Appearing in the places where heavy machinery has devastated the soil and native habitat, Scotch broom grows and helps to heal the earth with nitrogen-fixing roots, as do many of the plants in the Fabaceae or legume family." Boyer notes the plant's ability to combat erosion and says that in certain Northwest ecosystems, it has been observed to die back naturally after soil amendment or when eroded landscapes have stabilized: "[When] local trees of the area return and get larger, Scotch broom eventually gets crowded out and dies back." Other Northwest ecosystems, such as meadows and open prairies, are still at risk of being taken over by the industrious Scotch broom, especially since its nitrogen-fixing tendency is harmful to prairie plants that thrive in low-nitrogen soil. But as Boyer explains, "If practices such as controlled burnings were still utilized, this would not be such a problem."[23] While it is true that controlled burns would help this problem, it's simultaneously true that Scotch broom can produce 10,000 seeds *per plant* per year, each of which might lie dormant for thirty or more years.[24] It's undeniably a complicated equation. I believe human beings have a unique role in our world, as both conveyers of invasive species and mitigators of their environmental harm, but the indiscriminate bulldozing and poisoning of ecosystems, as in Grass Lake Nature Reserve, is not how I think we should be "managing" them. In fact, there are many studies that suggest that many accepted modes of invasive species eradication can have unintended and counterproductive ecological consequences.

A 2016 study titled "Potential Problems of Removing One Invasive Species at a Time: A Meta-Analysis of the Interactions Between Invasive Vertebrates and Unexpected Effects of Removal Programs" shows how the eradication of a single invasive species can cause harm to native species in an ecosystem, destroying the new balance that has been formed in that ecosystem. The paper's findings show that "the removal of a single invasive species always led to a negative or neutral mean effect on native species performance or survival" and "never found a positive effect size where the removal of one invasive led to an increase in native performance."[25] This is a lofty claim, certainly, but this was a peer-reviewed study and one that exposed the fundamental problems with hasty intervention.

As I said before, the total eradication method of invasive species management could only make sense on a small island not frequented by human activity—though, even then, the eradication of invasives has potential to negatively disrupt native ecosystems. In the story of the Judas goats on the Galápagos Islands, eradication of the invasive goat species led to the population decline of a native hawk species that had learned to rely on the goats for food.[26] Regardless, as with all invasive species eradication efforts, we need only count the days until the same or some other "invader" comes back. Hawaiian ecologist Don Drake remarked, "Until recently, everyone looked at native and non-native species as the good guys and the bad guys. Many people in the conservation community are growing to accept the fact that what we're going to live with in the future is a mix of native and non-native species."[27] What these mainstream conservationists have been fighting for is a world that simply cannot be reclaimed, no matter how many times they bulldoze and poison that Scotch broom. The term "invasive," with its specious yet compelling militaristic undertones, is a form of sophistic rhetoric that can easily shut down alternate perspectives and ultimately leads us into eradication efforts that are often futile or might even cause more ecological damage than the invasive species themselves.

The logic of invasive species management echoes (or is echoed by) nationalistic and xenophobic inclinations which are easily drawn out by allusion to war. First, consider Adolf Hitler denouncing the mixing of races in his text *Mein Kampf*: "All great cultures of the past perished only because the originally creative race died out from blood poisoning." And: "The Germanic of the North American continent, who has remained pure and less intermixed, has become the master of that continent, he will remain so until he, too, falls victim to the shame of blood-mixing."[28] But even today, there are innumerable eerie parallels between how ecologists speak about invasive species and how racist politicians speak about immigrants. Here is Donald J. Trump describing immigrants to the United States at a rally in New Hampshire in 2023: "It's poisoning the blood of our country, it's so bad, and people are coming in with disease, people are coming in with every possible thing that you can have."[29]

Compare these with a *Science* article describing conservation efforts to protect red wolves from interbreeding with coyotes:

> Biologists have successfully prevented coyotes from destroying the genetic integrity of red wolves, one of the world's most endangered canines, a new study concludes. . . . The

main threat to the wolves comes from coyotes, which mate with the wolves to produce fertile "coywolf" hybrids. This genetic dilution is a recent problem; coyotes migrated into the region after the FWS [Fish and Wildlife Service] biologists reintroduced the wolf. In 1999, biologists started sterilizing coyotes to prevent hybridization. The sterilizations are "not to control the coyote population size [but rather] purely to keep their DNA from being passed to a red wolf offspring," says ecologist Eric Gese of Utah State University, Logan, a lead author of the new study and a member of the panel overseeing red wolf recovery efforts.[30]

I don't think it's a stretch to say there are similarities between racist claims that we need to fight immigration to protect the purity of Germanic or American blood and environmentalist claims that we must protect the pure blood of a preferred native species.[31]

Invasive Terminology Today

The discourse that now exists around invasive species management makes it difficult to conceive of any alternate solutions, both because it is infused with the urgent imagery of war and because it is liberally sprinkled with ambiguous terminology.

Even the term "invasive species" lacks scholarly agreement and has many synonyms. In an article aiming to establish new, neutral terminology, environmental researchers Robert Colautti and Hugh MacIsaac identified over thirty terms common in the field, many used interchangeably with the term "invasive."[32] These include "alien," "colonizing," "escaped," "exotic," "foreign," "immigrant," "imported," "nonnative," "naturalized," "nonindigenous," "temporary," "tramp," "transient," "transplanted," "transported," and "weedy." While some are obviously xenophobic, others seem to have hardly any meaningful distinctions between them. After analyzing the historical use of invasive terminology, Mark Davis argues that the current definition of "invasive species" is varied, poorly articulated, and differs across cultural, political, and regional borders.[33] In the text *Fifty Years of Invasion Ecology: The Legacy of Charles Elton*,

Davidson M. Richardson lists almost nine pages of invasive species terms, explaining that his motive for establishing this list grew out of the varied and uncritical use of terminology in invasion biology. This uncritical use of terminology, he argues, serves to impede developments in the field.[34] Invasive species management must reckon with the lack of shared understanding of any of these terms. Despite individual attempts at clarification, invasive species management is a discipline still rife with contradictions and unscientific claims reinforced by militaristic metaphors, jargon, and plastic words.

Consider that both climate scientists and invasive species conservationists share a frantic urgency in their work. Yet climate scientists are fighting not only to be heard but also for their proposals to be seriously adopted. Meanwhile, we can witness policy after policy promoting invasive species management. One discipline offers tactics that do not challenge—and even thrive—under a capitalist agenda; the other requires us to fundamentally alter our society. At the end of the day, without abandoning capitalism, we will only see biodiversity loss accelerate. A post-capitalist world that prioritizes people and planet is the only real solution to the invasive species problem. Until then, invasive species management will continue to be a pursuit both Sisyphean and profitable.

Plastic Words in Invasive Species Management

It can be no surprise that Uwe Poerksen's omnipresent plastic words run rampant in the discourse around invasive species management. In 2018, I studied how often plastic words show up in six invasive species documents designed as an information campaign targeting the general public and easily available on the National Invasive Species Information Center website. I was able to reveal a statistically significant presence of plastic words at 2 to 10 percent of these documents. This may not seem like a lot until you compare that with the frequency of prepositions, such as "in," "at," "by," "on," "with," and "to," which are fundamental to sentence structure. The *Chicago Manual of Style* recommends the use of one preposition for every ten to fifteen words, or between 6 and 10 percent of the text. The fact that vague plastic words can appear as often as prepositions in these documents indicates an almost fundamental reliance on them in the present discourse of invasive species management.

Alongside this research, I also conducted a survey to determine the degree of readability of one of those documents, selecting one that contained plastic words as roughly 10 percent of the text. This document was designed to clarify confusing definitions of "invasive species." Ironically, the document did not score well in terms of readability, with survey respondents remarking on the same words being used in variable ways. The results revealed a telling trend—the same parts of the document would be considered equally clear and unclear by different survey participants. This seemingly contradictory response is characteristic of plastic words: they are familiar enough that most people think they know what they mean yet find them difficult to define when pressed. Poerksen says of plastic words, "When they first appear, they are fashionable and command attention; but then they merge with the everyday and soon seem commonsense."[35] This is their trick. They settle snugly into our vernacular, and we assume we know what they mean.

One survey participant wrote, "Risk/benefit assessment, regulatory, invasive: none of the vocabulary words were very difficult for me to understand, but these could possibly have multiple contexts/meanings." While these are not plastic words, as each has a specific definition, the observation nevertheless reflects another fundamental attribute of plastic words. As Poerksen observes, "The effects radiate into language."[36] Poerksen identifies how "usage of these words colors their environment: in their vicinity, other words and word groups are taken up and redirected."[37] It seems that when plastic words are used, their ambiguity will bleed into the rest of what is being said, and in this manner, plastic words colonize language. Here again, we can see a similarity between how plastic words act in their environments and how we are told invasive species act in new ecosystems.

Plastic words, which can lend any message dubious intellectual authority, can also render the rest of that message hollow by association. It is difficult to ascertain meaning when language is structured in relation to meaningless plastic words. By virtue of words such as "development," "program," and "management," a whole document is drained of linguistic precision. Therefore, the count of plastic words per document is less important than their mere presence. From this, we can see how the plasticity of language is part of a broader trend of inaccessible communication. Simply eliminating these terms will not necessarily clarify meaning, as plastic words can turn the language they neighbor into a purgatory of confusion.

The terms associated with invasive species ecology are variable—many different words may mean the same thing, while a single word may mean many things. "Invasive species" is a phrase that replaces more specific terminology and is a term often used uncritically. It is metaphorical, linking niche elements of scientific study with more commonly comprehensible phenomena. In this way, the phrase "invasive species" acts much like a plastic word. It is not truly plastic, but it carries many similar characteristics, of which it is wise to be wary.

From here, we must unpack the technical definition of "invasive species" itself to finally deflate this outdated concept. We know now *how* the story of the invasive species is being told. It is time to turn to the question of *why* this singular perspective persists.

The Problem of the Invasive Species Story

According to the National Invasive Species Council, an entity tasked with clarifying what constitutes invasive species in the United States, invasive species are defined as species nonnative (or alien) to the ecosystem under consideration and whose introduction causes or is likely to cause economic or environmental harm or harm to human health.[38] While this definition might appear straightforward, it is worth unpacking.

Many ecologists posit that nativity is assigned based on whether a species has been introduced to a new habitat by human intervention (nonnative) or natural migration (native). Still, determining the true nativity of a species is extremely difficult (if not impossible), since historical data on nativity is fragmented and brief. Many species we consider native in fact have their origins elsewhere, like honeybees and earthworms in North America. The further back in time we go, the less relevant the term "native" becomes, as nonnative species naturalize. Additionally, as anthropogenic climate crisis continues to accelerate and alter weather and temperature patterns, the incongruity of nativity is further exacerbated.

In the 2011 article "The Rise and Fall of Biotic Nativeness: A Historical Perspective," coauthors Matthew K. Chew and Andrew Hamilton argue that the concept of nativity in invasion biology is a weak, contingent, and variable paradigm that leans heavily on subjective human bias. While many ecologists recognize the natural migration of species into new ecosystems as natural activity, nativity still often operates as the ideal that presupposes that a species belongs in a singular locale. More accurately, species don't belong to a place but instead are part of a vast network of relationships, entailing weather and temperature, soil acidity, and other plants and animals and fungi.[39] Fixating on belonging requires ignoring the ways species actively create their ecosystems. Habitats are not passively awaiting organisms: organisms organize habitats.

In his book *Biology as Ideology*, Richard Lewontin argues that, unlike the theories of Darwinism, which suggest this one-way street of organisms as passive agents shaped by their active environments, organisms that thrive in habitats act on their environments as much if not more than their environments act on them. "Organisms," he explains, "do not experience environments. They create them. They construct their own environments out of the bits and pieces of the physical and biological world, and they do so by their own activities."[40]

Climate Crisis and Climate Refugee Species

During climate crisis, the concept of nativity is made more complicated when we consider how climate refugee species are forced to leave their old uninhabitable habitats or go extinct.

In May 2016, on a break from work in downtown Olympia, I joined a small crowd gathered on a wharf to observe about half a dozen dolphins that had entered Budd Inlet. I later spoke to a local who said this was the first time the common dolphin (*Delphinus delphis*) had been seen anywhere in the Puget Sound in recorded history. Somehow, they had made it all the way to the southern tip of the inland waters, far from regions to the south, which were getting too warm.[41]

As climate journalist Eric Holthaus writes, "Species that are migrating due to climate change don't respect the borders of nations or national parks. 'The old ways of conservation are entirely out the window,' [Australian marine ecologist Gretta Pecl] told me. 'We used to want to protect things in little particular areas, but things are not the same and never will be the same. We have to actually have some kind of large-scale strategic way of thinking about how to manage species.'"[42] Mark Davis likewise argues for the reconsideration of what constitutes an invasive species in light of climate crisis. Historically, the movement of many invasives and nonnative species has been the direct result of human activity. Today, the human-caused climate crisis also spurs the migration of species into new areas.[43] Is this considered natural migration because the species are moving themselves, or is it human intervention because anthropogenic climate crisis has forced species to flee their traditional habitats? There are ecological and ethical considerations that must be taken into serious account when considering the eradication of nonnative or invasive species, especially because an estimated 25 to 85 percent of species are seeking new habitats due to climate change.[44] As journalist Marina Bolotnikova puts it, "Climate change and the range shifts it's causing are extraordinary circumstances. If a species flees a habitat that is burning or melting, is it ever fair to call it invasive? Even outside of a climate context, this tension reflects a more fundamental problem within the invasive species paradigm. If the label is so stigmatizing that the only appropriate response feels like extermination, perhaps something else needs to take its place."[45] Would we bulldoze and spray the entire world because plants and animals are on the move?

Davis also alludes to how our changing climate will distort the static distinction between *native* and *nonnative* and to the possibility of native species acting in ecologically harmful ways. He cites the case of the mountain pine beetle, which is native to the United States but has become an epidemic particularly in the Rocky Mountain region. Here a native species acts invasively—and is being treated, managed, and referred to as an invasive species—but does not fit the technical definition of "invasive." Conversely, climate crisis enables some previously invasive species to become less harmful, such as the larch bud moth, which has exhibited ecologically problematic outbreaks historically in southern Switzerland except between the years 1981 and 2007, when no outbreak was recorded. Studies suggest that it is warmer weather that has decreased these problematic outbreaks. This suggests that invasiveness is a contingent rather than an inherent, ontological category, subject to changes in definition as climate change continues to alter ecosystems.[46]

Life on earth is in a process of radical movement and reorganization in response to global capital and the anthropogenic climate crisis. Attempts to prevent or inhibit this movement of species migration are both hubristic and doomed to fail because life itself *is* movement and, under climate crisis, that movement is rapid and essential for the preservation of displaced climate refugee species. Invasive species management is alleged to preserve biodiversity, but, under climate crisis, the impulse to eradicate "invasive" climate refugee species might instead endorse extinction.

Where and why does a given species belong to a specific place? Does a species belong solely to the location from which it first evolved? Nativity is an old ideal that frequently sets a lofty conservation goal of restoring ecosystems to how to pre-colonial status. But as Chew and Hamilton explain, "What 'human intervention' means to each depends on cultural context. . . . In the Americas, 'native' usually identifies taxa believed present before 1492, even though there is good evidence of widespread, long-distance trade and plant domestication by pre-Columbian Americans."[47] It is a ultimately an ideal born out of the belief in the separation of humans from nature, one that is inherently foundational yet deceptive, and which ignores the ways Native people have altered natural habitat since time immemorial. With their myth of origins, those who uphold these puritanical ideas around nativity aspire (sometimes unwittingly, sometimes intentionally) to fix and contain a volatile nature—a living world that has always been shifting, changing, adapting, and moving.

What Does "Harm" Mean?

Now, on to the second part of the definition of "invasive species": "An invasive species must cause or be likely to cause economic or environmental harm or harm to human health." This part of the definition offers a few more complications than the first part. In order to be labeled invasive, a species can certainly be extremely destructive ("harmful"), like the Asian chestnut blight fungus, which swept across thirty-five thousand square miles of the eastern United States in the nineteenth century, killing almost all chestnut trees in its path.[48] But invasiveness, according to this definition, appears to cast a wider net.

Right off the bat, determining the possibility of causing harm, especially harm to human health, is difficult and subjective. "Harm" and "health" are plastic words, and subjective concepts that can mean different things to different people. The National Invasive Species Management Plan recognizes this variability of the word "harm," and the "Invasive Species Definition Clarification and Guidance" white paper points out that while some people may consider a species harmful, others may recognize and reap benefits from that same species.[49]

Moreover, there is a dose of geographic alchemy at play, as a species declared invasive in one region will be de facto condemned in another, far-distant region without clear reasoning. This creates a sort of myopia among invasive species managers. In Australia, non-native devil's claw (*Martynia annua*) has been the subject of costly ecological eradication efforts for over thirty years, where research has found no evidence to suggest that devil's claw contributes to biodiversity loss or interferes with nutrient cycling.[50]

Additionally, our cultural associations with species far outweigh these definitions. You may be the unwitting caregiver of an extremely ecologically devastating invasive species. Domestic cats are a non-native species that severely depletes wildlife species, such as birds, squirrels, rabbits, and other creatures. Studies suggest that cats have been responsible for the extinction of up to sixty-three bird, mammal, and reptile species.[51] Yet many of us (myself included) would be up in arms at the thought of a systematic conservationist massacre of our furry feline friends.[52]

The vagaries of definition are further compounded by the uncertain time frames over which a species may appear to be troublesome. A nonnative species can take many years before becoming invasive.[53]

Further, by destroying a nonnative species before it causes any harm, we lose the opportunity to examine its effects on an ecosystem's resilience or any possible long-term benefits of that nonnative species. A 2011 study helped to determine how nonnative and even invasive species sometimes work within their introduced ecosystems to form mutually beneficial relationships with native species. The researchers, Tomás Carlo and Jason Gelditsch, found that the honeysuckle plant (*lonicera*), which was invasive to a particular region, actually began to form mutually beneficial relationships with native bird populations and with other native plant species. There is also no escaping the fact that when native species decline, the ecosystem may benefit from introduced species that serve to fill the roles of those decreasing native species. Carlo argues that, given time, some invasive species may seamlessly slip into near-empty or damaged niches to help fulfill ecological functions and, in so doing, help rebuild and strengthen biodiversity in an organic, self-sufficient way.[54] "'Nature is in a constant state of flux, always shifting and readjusting as new relationships form between species, and not all of these relationships are bad just because they are novel or created by humans,' Carlo said. 'We need to be more careful about shooting first and asking questions later—assuming that introduced species are inherently harmful. We should be asking: Are we responding to real threats to nature or to our cultural perception and scientific bias?'"[55] With enough time, might certain invasives integrate within local communities and prove beneficial as the ecosystem adjusts? Sometimes destroying an invasive species in a given habitat is counterproductive and serves to disrupt the "newly formed balance of an ecosystem."[56]

The harm caused by invasive species can be hierarchical, and the first type of harm described in our definition of invasive species is economic harm. According to this view, the only sin of an invasive species may be that it can decrease property value; some invasive species are less harmful than they are supposedly ugly. Vandana Shiva, in her *Monocultures of the Mind: Perspectives on Biodiversity and Biotechnology*, describes a classic example of how dominant knowledge systems can discount native plants as weeds if they are not economically viable to those systems. An international forestry consultant speaking about an ecosystem of the humid tropics is quoted as stating, "The important question is how much of this biomass represents trees and parts of trees of preferred species that can be profitably marketed. . . . By today's utilization standards, most of the trees in these humid tropical forests are, from an industrial materials standpoint, clearly weeds."[57]

Shiva illustrates an example in which the plant bathua (*Chenopodium album*), native to India, is nevertheless considered a "weed" by dominant knowledge systems. Bathua is nutritious and especially rich in Vitamin A but poses a threat to the growth of more commercially viable wheat. Wheat has a high market value internationally, while bathua is prized locally. Because bathua threatens the productivity of wheat, it is "managed" with herbicides. These herbicides, however, also kill the wild reeds and grasses that rural Indian women use to make baskets and mats—a source of income for rural communities. Meanwhile, forty thousand children go blind in India each year because they do not have enough Vitamin A in their diets. When dominant knowledge systems prioritize market value and monetary gain, local ecological and agricultural knowledge is negated if those plants threaten commercial gain.[58]

Another, more widespread example—cattle. Cattle domestication can be traced back to Southwest Asia, from where it spread initially to the Middle East, India, and Europe before becoming the center of a worldwide industry.[59] Cattle could therefore be defined as "nonnative" to many parts of the globe. The industry of cattle farming is also extremely ecologically damaging in most places, causing polluted runoff, releasing methane and nitrous oxide, overgrazing, and in multiple ways destroying ecosystems and decreasing biodiversity, making cattle not just nonnative but also very much an invasive species across much of the planet.[60] In some instances, native prairie ecosystems are obliterated and converted into grazing fields designed specifically for cattle.[61] In other instances, native plants that are toxic to cattle, like larkspur, are "managed" by way of chemical herbicides.[62] Moreover, studies suggest that raising cattle for food consumption (milk and beef) is one of the leading causes of the present climate crisis.[63] Despite this impressive listing of ecological harm, cattle are *not* considered invasive but instead are subsidized by many governments due to their economic prospects.[64]

Under capitalism, agribusiness is paid to raise and breed one of the most destructive invasive species on our planet. This offers quite a different way of understanding "invasive species management," where "management" here means the subsidized growth rather than the absolute eradication of an invasive species.

Shiva writes, "The one-dimensional perspective of dominant knowledge is rooted in the intimate links of modern science with the market."[65] Harm to capital and the economy is explicitly part of the definition of "invasive species." In invasive species management, as in many disciplines in our capitalist society, we are inclined to think with our wallets. Environmentalist and philosopher T. J. Demos writes that perhaps we should "ecologize the economy rather than economize the environment."[66] Technically, a species needs only to have the *potential to* decrease the value of a property *on another continent* to be labeled as invasive. This is used to warrant costly prevention and even ecocidal eradication measures in favor of potentially preserving banks. Per these capitalist priorities, the endeavor to eradicate invasives can have far more detrimental environmental consequences and harm to human health than the invasive species themselves.

Invasives in Cop City

Let's look at an example of when the presence of invasive species in a particular American forest not only spelled doom to an entire thriving ecosystem but also resulted in the death of a twenty-six-year-old.[67]

In Atlanta, Georgia, there is a dense and marshy forest called the South River Forest or the Weelaunee Forest.[68] In 2021, the city government tried to secretly assign 300 acres of the forest to the Atlanta Police Department and 170 acres of neighboring forest in Intrenchment Creek Park to the film production company Shadowbox Studios. The intention was to bulldoze the forest and replace it with a gigantic police training compound (complete with a mock city inside it), the country's loudest soundstage, and an airport.

After the public learned about this development project, Atlantans displayed an upsurge of concern that culminated in a multiyear forest defense movement known as "Stop Cop City" or "Defend the Atlanta Forest." Besides fury at the promise of increased police presence in the surrounding community, which is predominantly Black and has strained relationships with the police due to their tendency toward racism and violence, there are also substantial environmental concerns. This forest has been referred to as "the lungs of the city" and provides ecological protection in the form of capturing carbon, filtering toxic runoff, and preventing flooding. It is a living ecosystem and is habitat and home to many different species. Migratory birds flock there, and it is one of the only breeding grounds for local amphibians. Defenders of the forest write: "Our canopy is the main factor in ensuring Atlanta's resiliency in the face of climate change."[69]

Dustin Hillis, chair of the Atlanta City Council's Public Safety Committee, has stated, "It isn't what I see as a 'forest,' especially not an old-growth forest. If you go out to the land, which I have been to many times, there's a lot of invasive species."[70] Here we see how an entire viable and fundamental ecosystem is being threatened for being home to a handful of invasive species. This is the logic of Atlanta City Council: since there are too many invasives, the ecosystem is essentially illegitimate. They maintain that, because of the invasive species, it is a landscape both devastated and degraded, and this is the argument they use to justify their desire to invade it and further degrade and devastate it by way of development. It's absolutely ironic, and this mentality can also be understood as a metonym for capital's deprivation of the entire planet's ecosystem. Since there are invasive species everywhere on earth, nowhere on earth is worth saving. So, power up your excavators, boys; we might as well develop!

Despite what Hillis said, Weelaunee Forest—even if it is home to invasives—is still a forest. It is a vibrant ecosystem providing crucial habitat for native species, wetland filtration, protection from flooding for neighboring human communities, and much more, and it has been primarily condemned due to the alleged presence of invasives. The generally understood argument against all invasive species can often excite errant environmentalists concerned about their favorite park—until they realize that the cost of eradication means waging an eternal and absolute war on that park and destroying their nature trails. After all, many accepted modes of eradication can have unintended and contradictory consequences—herbicides do not differentiate between invasive and native and will poison without prejudice. Yet the justification to poison or destroy an entire ecosystem based solely on the presence of invasive species will often go uncontested.

In Atlanta, protests, parties, and resistance under the battle cry of "Stop Cop City" surged over several years but to little to no avail, despite mass national support. Thanks to the misguided, oversimplified condemnation of invasive species, we were left with dozens of forest defenders facing felonies, the promise of an obliterated ecosystem, and even the Georgia State Patrol killing of Tortuguita, an unarmed forest defender. In this forest, it is not the invasive species that are causing environmental harm or harm to human health—rather, here is a clear example of when the intent to eradicate invasive species is far more harmful than the invasive species themselves. Here, the cure is worse than the disease. While I do not believe in the prominent narrative surrounding invasive species management, I concede that there are indeed extremely harmful invaders in this Atlanta forest. But these invaders are not plants, animals, nor insects, nor fungi. They are not loblolly pine, box elder, Chinese privet, Callery pear, tree of heaven, chinaberry, or kudzu. The real invaders in this story, the ones causing the most harm, violence, and ecological devastation, are the Atlanta City Council, Shadowbox Studios, and the Atlanta Police Department.

Are We Invasive Species?

What are the boundaries and limits of what constitutes an ecosystem, a forest, or an invasive species?

Let's reflect on ourselves. Our narrative is interwoven with the narrative of invasive species—they are not an unnatural other, and their omnipresence is absolutely and entirely anthropogenic. Aside from the fact that most invasive species have invaded ecosystems because human colonizers have brought them there, is it surprising to recognize that many humans themselves very nearly fit the technical definition of invasive species?[71]

We are a species originating from a single region but have since spread globally. And, more recently, global capitalism has essentially required nearly every person on this planet to contribute to "economic harm, environmental harm, and harm to human health" on a daily basis. Let's look at our unfortunate ability to change the very climate of our world. According to the 2016 Environmental Protection Agency report *Climate Change Indicators in the United States*, the global burning of fossil fuels and deforestation for the sake of development, power, and profit are the main contributors to greenhouse gases and climate crisis, which in turn destroy human and nonhuman communities, kill billions of animals and around five million people annually, and promise an increasingly unrecognizable and uninhabitable planet.[72]

Biologist Brenden Larson muses, "We may dislike IS [invasive species] because we observe something in their behavior that we dislike about our own." He notes our similar tendencies to expand and spread and argues that "we characterize IS as amoral in terms of numerous traits—aggressiveness and lack of control, in particular—that 'represent forbidden sides of human nature.'"[73] What we fear in them, what we dislike about their patterns of behavior and tendencies, may reflect what we fear and dislike in ourselves. But this way of living with the world is not our fault, nor is it unavoidable—many traditional Indigenous land stewardship practices show how people can intervene with nature in ways that do not cause harm. It is not that we are invasive species, but rather that we are forced to live under an invasive society.

Within the ordinary routines of capitalism, we have almost all been made to act as invasive species, but it is the people upholding this extractive and exploitative society who are truly to blame. This is a systemic problem with a systemic solution: if anyone is to take the concept of invasive species management and eradication seriously, it will first and foremost mean the eradication of a global capitalist society.

Invasive Species and Biodiversity Loss

By the early 2000s, it was estimated that the United States was spending $120 billion on invasive species per year, while today, the global cost (also per year) is about 423 billion.[74] These staggering figures are fueled by the widely accepted claim that invasive species are the second leading cause of biodiversity loss next to habitat destruction.[75] However, there is evidence that undermines this assertion.[76]

As quoted in Matthew Chew's "Ecologists, Environmentalists, Experts, and the Invasion of the 'Second Greatest Threat,'" botanist Ken Thompson has declared, "On reflection I think that assertion [that alien species constitute the second greatest global threat to biodiversity] has been debunked so often (yet is endlessly repeated) that it no longer deserves the status of a myth, and is best described merely as a straightforward lie."[77] Mark Davis points out a Canada-based study that "found introduced species to be the least important of the six categories analyzed (habitat loss, overexploitation, pollution, native species interactions, introduced species and natural causes)" as regards to imperiled species.[78] Researchers show that the data around biodiversity loss for the United States is skewed by the inclusion of information about Hawaii, which experiences more drastic biodiversity loss because it is an island.[79] A report published in 2004 entitled "Are Invasive Species a Major Cause of Extinctions?" notes that there is very little data to support the conclusion that invasives decrease biodiversity. In response to skepticism, the researchers of the report stood firm in their analysis, arguing:

> Recent work suggests that even if zebra mussels and other stressors were removed, these long-lived species would not recover because of the legacy of toxic sediments left from decades of pollution. We feel that it is essential to determine the most important factors impacting declining species; failing this, our efforts to save or conserve endangered species will be for nothing. If unionid populations were not already depressed, would invaders have the same impact? We argued that many threatened taxa are affected by multiple stressors, making assessment of the nature of the threats posed by invaders extremely challenging. Correctly assessing the crucial factors responsible for and threatening species or ESUs with extinction is ultimately vital for the preservation of biodiversity.[80]

The claim that invasive or introduced species are the second leading cause of biodiversity loss is clearly lacking nuance, and the reality is far more complicated. In other studies, the presence of invasive species has been shown to actually increase biodiversity in ecosystems. Ecology professor Dov Sax found that while invasive species sometimes decreased the number of birds on certain islands, plant species often doubled. "Species addition and biotic resistance experiments performed at local scales (generally less than [one square meter]) indicate that even species-rich plots that are more difficult to invade can, nevertheless, still be invaded; further, the net consequence of these invasions is generally an increase in total species richness." He goes on to explain, "Quantitative studies of invasions at regional spatial scales suggest that most regions of the world can absorb more species by invasion than they lose by extinction, such that the net effect of invasions in many cases is a large net increase in species numbers." Sax specifies that for "vascular plants, which generally show a pattern of few native species lost and many exotic species gained, invasions over the past few hundred years have led to a doubling of richness on many islands (e.g., from 2,000 to 4,000 species in New Zealand) and on the order of a 20% increase in many mainland regions, such as individual states in the USA." He adds, "For some other groups the increase in richness has been more extreme; in Hawaii freshwater fish richness has increased by 800% with the introduction of 40 exotic species and the loss of none of the five native species." Additionally, he cites Mark Davis to illustrate "that there is no evidence that competition from exotic plants has caused any native species extinctions, that is, the complete loss of a species from the planet."[81]

While the claim that "invasive" species can increase biodiversity seems counterintuitive, studies do back this up. However, as philosopher of science Christopher Hunter Lean points out, increased biodiversity does not necessarily imply a thriving ecosystem. Lean explains, "Local increase in species richness has been coupled with global species loss." This phenomenon has been described as "the biodiversity paradox." If a larger number of common, nonindigenous species move into an area, which simultaneously loses a smaller number of endemic or rare native species, we will see increasing local species counts yet global species loss—a loss in the total variety of species around the world. Lean quips, "Australia (and the world) has lost the desert bandicoot (*Perameles eremiana*) but gained the red fox, cat, black rat, and common pigeon; a triumph!"[82]

Still, as we learned from the research of Tomás Carlo and Jason Gelditsch, there are instances in which the presence of new invasive species does promote the overall health of habitat. A 2016 article titled "Biological Invasions, Ecological Resilience, and Adaptive Governance" by researcher Brian C. Chaffin highlights the ways invasives can strengthen and increase the resilience of ecosystems. Some invasive species have been known to help a depleted ecosystem recharge and gain strength by fulfilling functions that have been in some way stressed or degraded, as in the case of the invasive European green crab (*Carcinus maenas*) moving into the waters of the eastern United States. Overfishing of predators of a native crab, the purple marsh crab (*Sesarma reticulatum*), has enabled Sesarma to deplete the marshes. The presence of Carcinus, however, has contributed to predation of Sesarma, which has in turn helped restore native marshes to their previous ecological niche. An invasive species can also reinforce beneficial roles already present in an ecosystem through a process of redundancy. A study on the Florida Everglades found that native vertebrate faunas had experienced a decline, while invasive vertebrate faunas were spreading. Despite these changes, the ecosystem remained wholly unchanged. In these instances, the invasive fauna has reinforced certain functions that the native fauna contributes to, benefiting the ecosystem as native fauna become scarce.[83]

While it is undoubtedly a complex phenomenon, it has been clearly identified that in some cases, so-called invasive species can do more good than harm, while so-called invasive species management can do more harm than good.

If all this is true, why the unabating attack on invasive species? Why not reallocate all this money and these resources into more legitimate and lasting modes of conservation—like building more wildlife sanctuaries, creating wildlife bridges over highways, reducing pollution, or outlawing bioaccumulating pesticides? This fixation on invasive species is based on the unnuanced claim that invasive species are the second leading cause of biodiversity loss.[84] But what about the *leading* cause—anthropogenic habitat destruction? Capitalism, global trade, mass production, poison, and pollution are the leading causes of climate crisis and modern biodiversity loss. This includes the emissions and chemical runoff that come from manufacturing the pesticides used in invasive species management and the fossil fuel burned by trucks that bring those pesticides to wreak havoc on ecosystems, indiscriminately obliterating native and invasive species alike.

Rather than an effective tool for conservation and biodiversity, I argue, invasive species management is primarily a distraction, and invasive species are being used as scapegoats. This allows the public to believe that legitimate conservation efforts are happening while more severe contributors to biodiversity loss and environmental destruction are permitted to continue.

Invasive Species as Scapegoats

In Hawaii in 2023, wildfires broke out and practically destroyed the town of Lahaina on the island of Maui. An estimated one hundred people died, and twenty-two hundred structures were damaged or destroyed. What caused the fires were outdated, faulty, bare, and uninsulated power lines that were deemed a serious public risk and "an obsolete 1960s standard" by the power company themselves.[85] Somehow, however, the media pushed the story that the invasive grasses that grow around Lahaina were to blame. The fault was not identified as capitalist corner-cutting or the colonizers who introduced these ornamental grasses, but rather the grasses themselves.[86] These grasses are a scapegoat for much more systemic causes.

In the Pacific Northwest, where I live, alleged restoration efforts in the Olympic Mountains have included the elimination and relocation of mountain goats. Though native to some of the Pacific Northwest mountain ranges, like the Cascades, these mountain goats are not native enough to others and have been picked off, one by one, by the Washington Department of Fish and Wildlife. Some have been moved to the Cascade Mountain Range and others shot and killed by volunteers.[87] Meanwhile, in forests surrounding the Olympic Mountains, large-acreage clearcuts of native coniferous trees are so common that whole mountainsides resemble shabby buzzcuts, with random bald spots sprawling over the landscape. It should go without saying that clearcutting and planting monoculture forests (which are more prone to forest fires) cause far more devastation to the forest ecosystem than a few packs of mountain goats do.[88]

The Washington Department of Fish and Wildlife emulates a practiced magician. We see the magic trick of invasive species mitigation practices at work, while up their sleeve, they hide the environmental degradation occurring on behalf of the capitalist machine. Here, invasive species management is a violent and intricate spectacle, serving to distract us. Invasive species are judged and harshly sentenced, while colonialism and capitalism continue to wreak havoc not only on select ecosystems but also on the entire world. The unlucky goats of the Olympic Mountains and the Galápagos Islands can be understood literally as *scapegoats*—creatures incapable of defending themselves against the blame of ecological devastation when it is far exceeded by anthropogenic contributors. Like goat sacrifices to an archaic god, our scapegoat invasive species continue to be ritualistically sacrificed, not for the protection or favor of a deity but so that capitalism can continue unabated. Both ancient ritual sacrifices and modern invasive species management offer arresting performances that reinstate the power and importance of the class (whether priest or policymaker) performing the sacrifice, while diverting attention from the exploitative practices of the ruling class. What the scapegoat narrative of invasive species provides is deliciously irresistible: someone else to blame.

Politicians and fear-mongerers are similarly quick to blame immigrants for job shortages and economic troubles that often actually arise from the very policies for which the politicians themselves are responsible. Under Bill Clinton's presidency, the United States pushed Mexico to sign the North American Free Trade Agreement (NAFTA). With NAFTA, Mexico eliminated import tariffs on US food and other products, resulting in an influx of cheap American corn and other foodstuffs into Mexico, which spelled economic ruin for Mexican farmers and farmworkers, and many immigrated to the

United States in search of work.[89] But once in the United States, these immigrants are blamed for stealing jobs—while in reality, statistics show they actually create more jobs as they open new businesses and bolster the economy with wages spent on products and services and taxes they pay in their new country.[90] The shortage of jobs is a result of US economic and foreign policies—including increased trade with China, outsourcing production, mass incarceration, and the reliance on industrial robots to replace human workers, trends spanning from the 1970s through today.[91]

Who gets blamed and punished for an influx of migration, whether by humans or nonhuman species? Yet who is responsible for perpetuating mass poverty and biodiversity loss in the sixth mass extinction?

We can see here two parallel stories of scapegoating innocent survivors, distancing blame from its true source: a xenophobic, ecocidal, capitalist way of living. Agents of the state frequently blame others for the conditions they have themselves created.

Today, as climate crisis devastates the planet and globalization reveals the depth of connectivity between environmental and social systems, we are faced with the task of reevaluating many of our established practices of relating to the earth. The authors of "Biological Invasions, Ecological Resilience, and Adaptive Governance" advocate for a shift away from the management of individual *species* to a model based on whole ecosystems. They explain, "Gaining an understanding of biological invasions in terms of ecological resilience allows for the deliberate engagement with resilience-based approaches to governance . . . that can coordinate the management of invasive species at scales relevant to ecosystems, ecosystem function, and the provision of ecosystem services, instead of at anthropocentric scales such as political and jurisdictional boundaries."[92] Like Tomás Carlo and Jason Gelditsch, these authors argue that an awareness that ecosystems function on intricate temporal and spatial scales is necessary for invasive species mitigation. They believe that this alternative, holistic approach can be achieved through adaptive, decentralized governance in which communication between the public and institutions flourishes.[93] It is clear that we need to change the popularized understanding of invasive species.

This invasive species story, as it is widely understood today, is fraught. The predominant narrative—not only in the culture of invasive ecology and in popular methods of invasive species management but also in our larger global capitalist culture—tells us that poisoning, bulldozing, developing, and killing actually save lives. This story ultimately perpetuates a way of living that is killing all of us—whether we're invasive or not.

This critique of invasive species management and the language and rhetoric it uses to embolden proponents of its singular perspective is only one example of how plastic words and metaphors are used across disciplines of power to promote perspectives that are propagandist, inaccessible, and undemocratic.

But I aspire for us to chart a new path—one that allows us to reintegrate our society into the vast ecosystem of life in mutually beneficial ways. We need new ideas, new stories, and new *solutions* to evolve this paradigm toward one that is post-capitalist and life-affirming, to hold the true perpetrators of biodiversity loss accountable, and to save our dying planet.

CHAPTER 4

A New Story for Invasive Species

*We tell ourselves stories in order to live. . . . We live entirely,
especially if we are writers, by the imposition of the narrative line
upon disparate images, by the "ideas" with which we have learned
to freeze the phantasmagoria which is our actual experience.*
—Joan Didion

I spent my early twenties trying to figure out what I wanted to be when I grew up, and at twenty-four I decided to pursue a master's in environmental studies. I enrolled in my local college with the vague hope that this program would equip me with, if not the skills, then at least a diploma that could enable me to pursue the sort of work I believed would be most objectively meaningful as our world succumbed to climate disasters. My first class was on forest ecosystems, and it was a warm, sunny day in April when wildlife biologist Lowell Diller was scheduled to speak to my class about his work, specifically on the infamous owl versus owl controversy that continues to unfold to this day.

During his speech, Diller described how, from the 1960s through the 1990s, mounting evidence indicated that the spotted owl (*Strix occidentalis*) was dying out. The main culprit was clearly logging, which was destroying the forest ecosystem preferred by the species. Still, another culprit was the barred owl (*Strix varia*), an invasive species from the East Coast that had spread across the continent following the path of the white colonizers of North America. When the barred owl settled in the old-growth forests of the West, its population spread and infringed on spotted owl habitat. The barred owl ate what the spotted owl ate; it lived where the spotted owl lived; it even looked like the spotted owl—but it was better adapted for survival. It had a larger hunting range and could raise more owlets at a time than the spotted owl. So, as the barred owl increased in number, the spotted owl began to die out.

The spotted owl is a gorgeous bird, slightly smaller and rounder than the barred owl. The spotted owl's coat also has more white feathers, so she almost looks like a plump grandmother barred owl. Historically, the spotted owl flourished in the old-growth forests—these owls prefer their forests dense, old, and expansive. The dependency of the northern spotted owl on the old-growth forest is reciprocal, and spotted owls evolved to fit a specialist niche in its ecosystem to the degree that the spotted owl is referred to as an *indicator* species—"a species whose condition will indicate the health of the entire ecosystem."[1]

While the barred owl has its charismatic "Who-cooks-for-YOUUUU" hoot, the spotted owl's is more of a bark. In 1969, this distinct barking sound echoed through the trees of Willamette Forest within earshot of an undergrad, Eric Forsman, who arguably started the owl versus owl debacle that now rages, decades later. Forsman was twenty-one years old and working at a Forest Service guard station during fire season when he heard this barking cry. In an interview, Forsman says he initially thought it was a "weird-sounding dog" before realizing that he might be hearing the calls of the rare northern spotted owl.[2] He mimicked the call, and the spotted owls called back to him—and, after a few days of this interchange, one unexpectedly swooped down to get a better look at the startled biologist. Through this chance encounter, Forsman began his decades-long struggle to preserve the spotted owl.[3]

In the 1970s, environmentalism was trending among the American public, thanks largely to Rachel Carson's *A Silent Spring*, the publication of which can be considered the start of the environmental movement. Carson's groundbreaking work highlighted the negative effects of pesticides and herbicides and warned of a future in which species are unintentionally extinguished as a result of indiscriminate human intervention in the environment. The effect was an unprecedented uproar of American citizens against unchecked environmental interference.

In response to public outcry, Congress passed several laws in the 1970s that served to protect the environment from degradation, including the National Environmental Policy Act, the Clean Air Act, the Clean Water Act, and the Resource Conservation and Recovery Act. In 1973, Congress passed the Endangered Species Act, based on the argument that all species have ecological and biological value, which demands the conservation of species that are endangered and prohibits activities that threaten them. These acts put limitations on industries and required "private and public decision-makers to consider environmental costs in actions affecting the environment."[4] This environmental momentum reached the forests of the Pacific Northwest, and the future of the trees and the spotted owl became a subject of great debate throughout the seventies and eighties. The spotted owl represented the forest—literally as species indicator but also symbolically. To save the owl meant to save the forest. After the northern spotted owl made it onto the endangered species list in 1990, the media hype and environmentalist push to conserve its habitat grew.

Logging and the timber industry initially appeared to be enemy number one. Logging began in the Pacific Northwest shortly after the first colonizers settled here. Frank Shaw, a Washington territorial official, spoke of the Pacific Northwest landscape: "The question was, shall a great country with many resources be turned over to a few Indians to roam over and make a precarious living on, making no use of the soil for timber or other resources, or should it be turned over to the civilized man who could develop it in every direction and make it the abiding place of millions of white people instead of a few hundred Indians."[5] As part of the violent colonization and displacement of Indigenous peoples, the settlers took to the trees and the land with gusto and greed. A 1981 book on the subject, *They Tried to Cut It All*, suggests the blind consumptive mentality of the loggers during this boom.[6] The original Western forests are estimated to have ranged from twenty to seventy million acres, but 70 to 95 percent of that had been logged by 2003.[7] "The tiny Nisqually reservation," Richard Kluger writes in *The Bitter Waters of Medicine Creek*, "is a place where Elder Jack McCloud still laments the disappearance of the cedars and charges timber giant Weyerhaeuser with destroying 'a part of the one great sacredness. . . . Trees produce oxygen; companies produce poison.'"[8] These old-growth trees also made up the ecosystem and habitat of the spotted owl. Studies have shown that pairs of spotted owls in Northern California used up to 1,900 acres of old-growth; in Oregon, pairs on average ranged 2,264 acres; and in Washington, they have been known to range an average of 3,800 acres.[9] Logging old growth is fundamentally fatal to spotted owls because they are entirely dependent on these forests.

So, in the 1990s, logging was restricted near spotted owl sites. Western Washington judge William Dwyer put a temporary stop to timber sales on twenty-four million acres of spotted owl habitat in seventeen national forests across the Pacific Northwest.[10] Through the efforts of people like Eric Forsman and the environmental movement of the seventies, the logging of old-growth forests was finally regulated. For a time.

But the spotted owls kept dying. The logging had already disrupted far too much of the spotted owl's habitat, which, combined with the robust presence of the barred owl, further threatened the spotted owl's preservation. Or so the story goes.

In the late 2000s, Lowell Diller initiated an experiment to see if killing thirty-six hundred barred owls in the Northwest would help spotted owls escape extinction. By the time he came to speak to our class, he had already personally shot and killed over seventy barred owls. As he told us this story, he explained that every time he shot down a barred owl, he felt like what he was doing was wrong. The first time he shot one, he said, he was so overcome with emotion that he nearly couldn't pull the trigger. He had spent his whole career learning to revere raptors only to voluntarily become their executioner. Even after killing seventy barred owls, he was still morally conflicted.

An animal rights group, Friends of the Animals, was one of the first organizations to protest this experiment, and they did so through a lawsuit. Their director argued, "To go in and say we're going to kill thousands and thousands of barred owls, literally forever, I don't see that as being a solution."[11] But Diller explained that it was the best available solution to the problem. After all, it was sort of working in some locations. In a little over half of the areas of the forests where barred owls had been killed, the spotted owl populations began to stabilize.[12] Though two of the sites showed almost no change after the barred owls were shot, Diller argued that it was the best plan they had; what else could be done to save the spotted owl from extinction?

As I sat in my forest ecosystems class, this story and all its implications stuck with me. As Diller spoke, I empathized with his dilemma. He called it a sort of Sophie's choice—and it made him sick to do it. He mourned the extinction of life and yearned for the preservation of an endangered species. He was loath to shoot down the very creature he had spent his life protecting, but to protect one owl he would have to kill the other. Still, this state-managed approach to biodiversity preservation seemed counterintuitive and cruel, even to the man responsible for the experiment's design.

After all, current thinking argued that barred owls had only come to the Pacific Northwest through European settler colonialism. Barred owls originated in the eastern United States and, as the theory goes, could not cross the treeless expanse of the Great Plains or the boreal forests of Canada until after colonization. Historically, Indigenous peoples likely incidentally suppressed barred owl expansion through the controlled burning of brush and trees in the Great Plains and in the Canadian boreal forests. However, when European colonizers arrived in North America and swaggered across the continent, they disrupted this form of land stewardship not only by violently suppressing Indigenous traditional controlled burning but also with the extirpation of bison and beaver, who naturally suppressed the growth of large forests. Without bison and beaver, boreal forest spread in Canada and little tree islands and pockmarked the Great Plains. This created two routes for barred owls to move west.[13] In the story of the barred owl and the spotted owl, we have a history of tragedy, evolving from millennia of sustainable human coexistence to the colonization and genocide of

Indigenous peoples, to ecocidal industrial logging, to a last-ditch effort of mass barred owl slaughter to save another owl species that has been all but doomed by logging and colonization.[14]

Like so many other invasive species, barred owls are innocent victims of colonization. They are trying to survive in a ruthless anthropogenic world that coaxed their ancestors across a continent, destroyed their habitat, and then finally decided to sacrifice great numbers of their kind for a poor and partial interpretation of the "greater good." Barred owls are a highly mobile species engaging in their innate capacity for range expansion. These owls have no false agenda, as our ecologists have even in their attempts at preservation, playing recorded mating songs to coax the barred owls out for a clearer shot.

As of late 2025, it has been nearly forty years since the spotted owl was placed on an endangered species list and nearly twenty years since the barred owl was placed on a hit list. In July 2024, the US Fish and Wildlife Service passed a new proposal for the extermination of nearly five hundred thousand barred owls across Washington State. Over seventy-five organizations have argued against this proposal, arguing that it is immoral given its scale and that it is ultimately unlikely to save the spotted owl from extinction or stop the barred owl from expanding. In their objection letter, they argue:

> The timber industry financed the "studies" and fieldwork that are the impetus for this owl-killing plan as an attempt to distract from the industry's continued destruction of spotted owl habitats. While the Biden Administration reversed a portion of the decision, the Trump Administration in early 2021 attempted to allow logging on up to 3.4 million acres of mature forests. It seems far easier, as a political matter, to authorize the mass killing of barred owls than to provide enduring and consistent protections of key habitats for the animals where there is a major political and economic influencer pushing for an expansion of logging opportunities.[15]

It is worth noting that the late Lowell Diller's employer was Green Diamond, a timber industry business whose website proclaims them a "forest stewardship company"—one that logs approximately forty thousand acres each year.[16] Diller was hired from 1990 to 2014 as a senior biologist with Green Diamond, and after his death in early 2017 the company endowed the Lowell Diller Wildlife Scholarship. It is ironic that a man so dedicated to the preservation of wildlife would spend his life working for an industry that greatly endangers wildlife. It is likewise ironic that Green Diamond would honor the legacy of a man who maintained widespread recognition of the endangered spotted owl population. But, after all, it was thanks to Diller's experiments eradicating barred owls to support the spotted owl population that animosity toward logging was redirected at another source. During Diller's initial experimentation to see if slaughtering barred owls would improve spotted owl populations, Green Diamond was one of the first to secure permits to shoot barred owls:

> Gary Rynearson, the company's chief communications officer and forest policy director, said that more spotted owls does not mean Green Diamond will increase logging, but it does mean that current rates of logging can continue. Though logging companies have often been at odds with threatened species such as the northern spotted owl, the company is excited about the study's results, Rynearson said.
>
> "When you can protect and sustain a business and jobs and also conserve the northern spotted owl," he said, "why not do it?"[17]

Despite the temporary regulations on logging placed around spotted owl sites back in the 1990s, it appears that logging spotted owl habitats is once again permitted—as long as the logging industry also has a permit to shoot barred owls.

Could Diller not fathom any alternative solutions to save the spotted owl because to actually address the root cause of spotted owl decline would mean challenging his employer? It seems that Diller's legacy may be a successful scapegoat story that sacrifices the barred owl so the logging industry can go on clearcutting without protest.

The Motives Behind the Jargon

To better understand the relationship between loggers, spotted owls, and barred owls, I spent hours digging through hundreds of pages of scientific studies, reports, and federal plans and policies describing the measures in place for preserving the endangered spotted owl. Over the course of this research, I experienced firsthand the difficulty of trying to comprehend the jargon-studded technical papers backing up such high-stakes policies. However, I was ultimately able to conclude that many of these studies were biased and flawed. Biased because many were conducted by logging companies with an inherent motivation to keep logging. Flawed because these studies argue that barred owls are the biggest threat to spotted owl populations—a scientific claim that cannot be proven because we have no way to control for the impact historical logging has had on spotted owl populations, and one that ignores broader habitat loss as well as climate crisis.

With the ongoing diverse efforts to preserve the endangered spotted owl, you would think loggers would, at the very least, give these birds a wide berth. You would also think loggers would be forbidden to harm, disrupt, or kill them. However, reading through these studies, I discovered a plethora of chilling loopholes and authorizations that shows just how much logging is allowed to get away with. Logging is still being conducted in spotted owl sites—even sites that were designated "set-asides" or intentionally excluded from logging. A buffer of only 0.09 to 1.3 miles is required between spotted owl nesting sites and logging sites, which is totally inadequate for a bird that requires thousands of acres to roam for food. Additionally, these sites are only protected during the breeding season, leaving spotted owls defenseless for the rest of the year.[18] These documents make clear that logging is actively destroying spotted owl habitat across Washington, Oregon, and California, and displacing spotted owls.[19]

Moreover, because logging is fundamentally ecocidal, it will have casualties. In the logging plan that Green Diamond presented to US Fish and Wildlife to detail its intentions and the effects of logging in certain areas over a number of years, the company suggested it would "incidentally take" no more than a certain number of spotted owls.[20] However, on two separate occasions, in 2007 and then again in 2019, US Fish and Wildlife allowed Green Diamond to increase the number of spotted owls they could "incidentally take."[21] Now "incidentally take" can mean "to kill," but it can also mean more benign forms of disruption. A generous and optimistic reader might argue that these permits allow Green Diamond to simply pursue these owls, perhaps to scare them off so they are not harmed during the clearcutting. However, these agreements actually permit loggers to kill spotted owls—just as long as minimum preservation measures are undertaken. A Humboldt Redwood Company study allows for 20 percent of spotted owl pairs from an allotted 108 spotted owl habitat sites to be sacrificed for the sake of logging and industry.[22] And as we have seen from Green Diamond's authorization to incidentally take more and more spotted owls, we can imagine minimum preservation measures might be similarly altered for the benefit of logging companies.

Meanwhile, management agencies and loggers blame spotted owl population decline on almost anything other than logging, including wildfires and barred owls. Barred owls are claimed to be the "largest negative impact on northern spotted owls," according to a US Fish and Wildlife Service report.[23] However, recent research makes it clear that logging is, in many cases, the primary reason spotted owls are dying out. One study analyzed recently burned forests to identify "how many of 105 burned northern spotted owl territories were logged before and after wildfires over an 18-year period (from 2000–2017), along with those where barred owls were also present, to sort out what might be driving spotted owl nest site abandonment that federal agencies blame on wildfires." In this study, "researchers documented that logging affected 87% of severely burned northern spotted owl sites, with barred owls recorded at 22% of the burned-and-logged sites." Not only does this show that logging impacted spotted owl habitat in far greater measures than barred owls have, but it also shows that loggers are absolutely continuing to log spotted owl habitat. In fact, a coauthor of this study dryly notes that "the U.S. Fish and Wildlife Service issued permits to the Forest Service to kill spotted owls like it was handing out candy at Halloween."[24] This study ("Logging, Not Wildfire, Is Most Likely Driving Northern Spotted Owl Decline") cannot speak to all spotted owl habitats across the Pacific Northwest, but in combination with the ample documentation available to the public, in the form of plans, proposals, and reports from both logging companies and the federal government, it is nevertheless clear that logging continues to be extremely detrimental to the spotted owl population broadly and is likely to be the primary cause of their decline.

These studies, reports, federal plans, and policies underscore the problem of inaccessible language that obscures material findings of cause and effect and how we cannot rely on experts motivated by profit to make the right decision.[25] Behind these studies is a vague promise that the logging industry is trying on some level not to kill spotted owls (though maybe they will sometimes, accidentally) and that loggers aren't logging spotted owl sites (mostly). But the experts here are clearly misleading, as they reluctantly admit instances of their guilt. Members of the logging industry are also simply lying as they point fingers at every cause but logging to explain the demise of the spotted owl.

In light of the oncoming extermination of nearly half a million barred owls over the next thirty years, the seventy-five organizations opposing the plan insist that "non-lethal management actions to protect spotted owls and their habitats should be made the priority action."[26] Meanwhile, after the echoing crack of the gunshot and the muffled thump of another lifeless owl, you can hear the screaming chainsaws of loggers in the distance, continuing to fell acre after acre of forest.[27] In all the decades of hemming and hawing over whether we need to kill this owl to save that owl, we have once again been made to forget the true culprit of biodiversity loss.

What happens after every invasive species has been eradicated? Or what if every single species stayed exactly within the confines of its range dating back to, say, before the Columbian exchange, when species were transferred between Europe and the Americas after first contact? Or perhaps to before the beginning of the Industrial Revolution, which, it is argued, spurred the onset of our current anthropogenic climate crisis? Imagine that invasive species management protocols as we know them now were long ago enacted and rigorously enforced, preserving whichever environmental era experts decided deserved preservation. How much better off would we now be?

I am certain we would still be facing mass extinction. Neatly resolving the problem of invasive species such that each is sorted into its appropriate range would do nothing to avert the climate crisis we now face. The system of nation-states, chemical manufacturing, and resource-extractive industries that undergirds our invasive species management regime would churn along unchanged. Carbon emissions would continue to rise year after year, and our biodiversity crisis would continue virtually unabated. Every primary driver of biodiversity loss would continue. Extinction would burn through the biosphere like wildfire in a drought.

Our current invasive species story treats invasive species as a kind of malicious entity acting outside "natural law" and warring against a pristine, stable, innocent ecosystem. And this story is ethically, economically, ecologically, and practically problematic—and above all, it distracts from the true culprit of biodiversity loss: a global capitalist society. But we can, step by step, challenge this narrative and push for another way of thinking about and relating to "invasive" species, ourselves, and our shared planet.

More Accurate Metaphors: From Managers and Invaders to Drivers and Passengers

The first part of tackling the problem of the dominant narrative involves addressing the insipid and misleading militaristic metaphors in the literature, both popular and technical, of invasive species management. This singular vision of war and the prevailing mentality of absolute eradication is an oversimplification in which more holistic, nuanced, and life-affirming proposals are largely ignored. Although invasive species can certainly be harmful to various habitats, the fostered panic is overblown and unnuanced. Alternate perspectives are emerging.

There is a growing understanding that invasive species do not *drive* or instigate the degradation of an ecosystem but rather arrive in the wake of an already degraded ecosystem. A metaphor that speaks of "drivers" and "passengers" in habitat disruption has emerged.[28] For example, the grassland ecosystem of the San Francisco Bay Area suffered from the "invasion" of nonnative serpentine grasses that threatened the habitat of the native Bay checkerspot butterfly (*Euphydryas editha bayensis*). Yet, in 1999, biologist Stuart B. Weiss discovered that these nonnative grasses were able to slip into the ecosystem as a result of car and truck emissions.[29] These air pollutants began breaking down the habitat's "invasion barrier," a sort of defense mechanism that helps a species and its ecosystem protect against competing species. In this case, the low nitrogen in the soil supported the ecosystem but reduced or eliminated the viability of potential competing species. However, the chemical composition of car emissions caused the soil to become nitrogen-rich, priming the soil for nitrogen-loving grasses.[30] To focus conservation efforts solely on eradicating invasive species is to ignore how they got here in the first place, which is to ignore the root problem of biodiversity loss and habitat degradation.

Are invasive species to blame in this biodiversity loss and habitat degradation? Certainly, some have negatively impacted ecosystems, but *are they to blame?* The answer is far from simple—rather, there is an intricate dance between anthropogenic disturbance and invasive species. We should shift our question from asking what the impacts of invasive species are to asking instead what is *driving* these species to invade. The driver of this particular serpentine grass invasion of the Bay Area was ecosystem disturbance due to human activity, and the grass was simply an unwitting passenger. Through the use of a passenger–driver metaphor, we can better tease out the capitalist, anthropogenic factors influencing the movement of species into new ecosystems.

Today many state agencies, conservationists, land management practitioners, and the general public recognize that the main instigators of biodiversity loss are habitat loss, invasive species, climate change, environmental overexploitation, and pollution. Each of these problems tends to be met with solutions specific to each instigator.[31] But to compartmentalize in this way ultimately undermines conservation efforts, which, to succeed, must address root causes and see these problems as interrelated. After all, can we not trace each of these instigators of biodiversity loss back to the basic fact of a global society operating in a capitalistic, extractive, and consumptive way? Rather than focusing on each instigator in turn, eternally, what if we addressed the root problem first?

Looking back at history, there is an integral link between capital's relentless expansion and the spread of nonnative species. From marine shipping to transcontinental highways to intercontinental air travel, the global capitalist economy drives the ecosystem impact we attribute to invasive species by directly introducing these nonnative species, which are literally and figuratively passengers in this larger system. If invasives arrive in the wake, then we risk conflating an effect with a cause. With the disruption of global climactic patterns triggered by capitalism's relentless greenhouse gas emissions and immense extraction of natural resources, all ecosystems are ultimately disrupted.

In the case of spotted owl population numbers, barred owls are not the original cause of their decline. Westward colonizers drove the shift in species composition by first disrupting the habitat of the Great Plains and then relentlessly logging the forests of the Pacific Northwest. In this story, the colonizer as both settler and logger is the driver; the barred owl is the unsuspecting passenger, now hunted and accused of the crimes of the driver. As long as our driver continues to be profit-motivated, this road trip will always end in ecological catastrophe. The discordant us-versus-them framing implied by our warmongering metaphors falls apart when we remember how anthropogenic invasive species are.

Ultimately, this war metaphor is overly rigid, distracting, and inaccurate, and, perhaps most importantly, it reifies the belief that people are inherently separate from nature. In this separation, we find it easier to excuse ourselves from any responsibility for the environmental problems our way of life has caused. We compartmentalize. We blame the invasive species for being exactly where we put them. In blaming the barred owl for the death of the spotted owl, we blame nature for her own demise without addressing our hand in it.

The Holistic Framework

I think that to fundamentally change the militaristic narrative around invasive species will absolutely require shifting away from viewing ecosystems through a subject-object binary, which suggests that the passive object (the ecosystem) is being acted on by a subject (a species), and instead understanding ecosystems through a relational perspective—considering how species and ecosystems interact with one another. As I have mentioned, ecosystems are not passive backdrops waiting for the perfect species to come along and fill their niches—species influence habitat just as habitats influence species. Understanding this requires us to think about ecosystems and their species—whether invasive or not—in a far more holistic way.

Consider the idea that no species is innately invasive (or disruptive). Determining whether or not a species is disruptive in a given, discrete area requires studying the relationships among the species present in the area. This is why a certain nonnative species may thrive in happy coexistence in one ecosystem but act invasively in other areas. In this context, invasiveness isn't about nativity or origin but rather about how that species gets along with the environment. Similarly, even people causing ecological harm are not innately invasive—this invasiveness is a result of how we are made to interact with our world under capitalism. This perspective on how species relate to their environments offers us a chance to understand species and ecosystems not as distinct entities but as relational beings.

In the passenger–driver metaphor, the global reach and intensification of the capitalist system are the fundamental drivers, with the invasives carried along in the passenger seats. This helps us recognize that it is not the species that are the enemy but rather climate change via capitalism. Biologist Brendon Larson reflects on the way traditional Chinese medicine practitioners think about diseases in terms of balance and harmony (even encouraging being kind to infections), as compared to Western medicine's militaristic approach to disease. Larson's holistic allegory of invasion ecology would better equip us to recognize the human agency in the spread of invasive species. Moreover, frameworks like these would offer us a chance to reintegrate ourselves into the natural world.[32] Indigenous Elders in the Kimberley region of Australia refer to native and non-native species acting discordantly with their habitat as "cheeky," a phrase that has both good and bad connotations, of being too clever or unchecked. Likewise, they describe an ecosystem as "healthy" or "sick" and speak of "looking after" or "caring for" the land.[33] English environmentalist James Lovelock put forward the Gaia hypothesis, which suggests that each organism and interwoven habitat in the world creates a self-regulating and resilient global ecosystem.[34] The late environmental activist Joanna Macy described a fourfold analytic framework by which we view the world: as a battlefield, as a trap, as a lover, or as the self.[35] These frameworks enable us to more usefully conceive of ourselves and our world.

Viewing our ecosystem as a holistic entity encourages both public discourse and natural resource agencies to break out of an obsolete way of thinking about ecosystems as rigid, stable structures that must be maintained and that crumble when change occurs. Instead, we must recognize them as living entities in constant flux. The "management plan" of killing half a million barred owls in the Pacific Northwest will not and cannot save the spotted owls. If these endangered owls don't perish from logging of their habitat or "barred owl invasion," they will certainly perish from unchecked climate crisis. Trying to preserve the spotted owl without addressing logging and climate crisis is pointless, as is trying to address climate crisis without addressing capitalism.

Above all we must recognize that people and nature are not separate—our fates are entangled, and more often than not we are creating the invaders that threaten our world. The sooner we rewrite the narrative accordingly, the sooner we can begin to find nuanced, holistic, and successful solutions to biodiversity loss.

Changing the Inaccessible Language

We have seen how jargon and plastic words used in both the scientific and public narratives of invasive species biology reduce comprehension and, with it, the possibility of finding communal solutions to the problem of biodiversity loss. We are then forced to rely on and trust a select group of experts to do what is right and to know what is best.

In the United States, state agencies such as the Environmental Protection Agency and the Department of Natural Resources and nongovernmental organizations like the Nature Conservancy focus on the preservation of habitat, the environment, and the natural world. But these institutions are run by unelected bureaucrat experts working in industries funded and overseen by a government dependent on a capitalist society. When only a few individuals can answer crucial questions or make consequential decisions, this indicates a severe restriction of the degree to which a society can be truly participatory or egalitarian. Perhaps most importantly, formal decisions on conservation and preservation policy are generally conducted at federal and state levels. These experts are beholden to organizations that owe their very existence to capital; profit (in one form or another) is their source of funding and power. The Washington Department of Natural Resources (DNR), for example, has this vision statement: "Our actions ensure a future where Washington's lands, waters, and communities thrive." Its stated mission: "Manage, sustain, and protect the health and productivity of Washington's lands and waters to meet the needs of present and future generations."[36] Meanwhile, the department has "a continuous flow of revenue" from logging and leasing land for agriculture, "communication sites, mining and mineral leases, wind farms and energy production, commercial properties, and rights of way."[37] As many of these involve substantial environmental damage, the DNR's funding sources are often fundamentally at odds with its stated aims. Meanwhile, a quick glance at the DNR annual reports show that the revenue generated by logging and timber sales alone accounts for around half or more of the department's income.[38] Even as the employees of the DNR may seek to preserve habitat, they are still at the mercy of the whims of funding and thus of capitalism, which is antithetical to environmental preservation. As Jan Matthews writes in *An Introduction to the Situationists*:

> As Marx pointed out, class society depends on the division of labor inaugurated through the division of mental and physical

labor. Capitalism further expands this division of labor by creating the need for the management and control of ever greater domains of social life. Capitalism produces a whole array of specialists (psychologists, professors, scientists, etc.) who work to perpetuate capitalism. We usually don't choose to be dependent on specialists, it is just the way the system is set up. A good example of this is the rule of specialists called politicians who represent people whether or not they wish to be represented. . . . "In a communist society there are no painters but at most people who engage in painting among other activities," wrote Marx.[39]

We are taught that redressing our ecological and social problems is best left to experts and their bureaucratic institutions that loom over us. But we cannot blindly trust experts funded by institutions beholden to capitalist accumulation and state power. They will always be in fundamental opposition to healthy people and planet, regardless of their stated aims and achievements. Marx's critique of specialization was advanced in the 1960s by the Situationists, an international revolutionary group based in France and made up of artists, writers, and political critics. In one of the movement's most famous texts, *The Society of the Spectacle*, author Guy Debord writes: "The root of the spectacle is that oldest of all social specializations, the specialization of power. The spectacle plays the specialized role of speaking in the name of all the other activities. It is hierarchical society's ambassador to itself, delivering its official messages at a court where no one else is allowed to speak. The most modern aspect of the spectacle is thus also the most archaic."[40] The hierarchy of power and decision-making embedded in capitalist specialization is a death sentence for our world. We must break away from the idea that only a select, biased few can decide for all; this comes from the corrosive philosophy of hierarchy and of the idea that everyone must compete to survive.

One possible way of incorporating a broader range of perspectives in scientific research is by using and prioritizing citizen science more frequently. Citizen science is "the practice of public participation and collaboration in scientific research to increase scientific knowledge. Through citizen science, people share and contribute to data monitoring and collection programs." In projects involving citizen science, scientists and researchers may recruit volunteers for data collection. Likewise, community-led groups may call on scientists and researchers to provide support for their own scientific projects. Some current citizen science projects include cataloging an array of species, documenting the bleaching of coral reefs, observing light pollution, and monitoring water quality.[41] CitizenScience.gov, a website dedicated to organizing opportunities in the United States for collaboration between scientists and lay researchers through citizen science, declares, "We believe participatory forms of discovery lead to better scientific outcomes and increase trust in the scientific process."[42] Through citizen science, the public can help guide scientific research more collaboratively and transparently.

By All Means, Call Out That Word!

In 2018, after defending my master's thesis, a classmate asked me about a word they needed to use in their own thesis.

"It seems like a plastic word," they told me. "I don't think I can use any other word in its place. What should I do if I have to use this plastic word?" They were essentially seeking guidance on their responsibility as a scientist in using these sorts of confusing, slippery words. I replied that while I thought it best, in general, to avoid these sorts of words, in many cases they were unavoidable. In those circumstances, I explained, I believe it's important to call out such oblique words that seem to avoid concrete definitions.

Uwe Poerksen has recommended adding a component to the part of dictionary definitions where a word's slang or archaic use is noted in italics. He suggests that we might add its plastic use too.[43] While helping the general public become aware of how plastic words operate, this might not necessarily be directly helpful to the wayward scientists who, stuck between a rock and a hard place, do not want to resort to such vague terms yet cannot use any other without some sacrifice of scientific integrity. Here I offer my own practical suggestion for scientists who must use this kind of language according to the standards of their discipline but who want to minimize inaccessibility, obfuscation, and impenetrability.

How to call out a plastic word: 1) name it, 2) define it in the context of how it is being used in your specific paper, and 3) bring attention to the fact that its definitions are variable and can change from context to context. For example, if I must use the terms "communication" and "environmentally harmful species" in a document, I might write this near the top:

Definition of Key Terms
There are several terms I use in this paper that I wish to call attention to because they are used across many disciplines and so carry different connotations, meanings, or associations.

Communication: Used in the context of this document, "communication" means the act of speaking, writing, or otherwise using language to convey thoughts and ideas. I refer only to communication between human beings.

Environmentally harmful species: The common definition of "invasive species" states that these are nonnative species that have the propensity to cause harm, including economic harm. Because harm can be subjective and differ from person to person (or ecosystem to ecosystem), because as an environmentalist I am not concerned with economic harm, and because native species can cause environmental harm, I use the term "environmentally harmful species" to describe what may have been formerly understood to be invasive species.

In the opening chapter, I explained my choice of the word "capitalism" to describe a particular phenomenon—and also why that term falls short. That was my callout—my preemptive clarification in an attempt to avoid misunderstanding throughout.

Many scientific papers will define specific terms they use to convey specific meanings. However, these papers rarely call attention to the varibale use of those words in other contexts. Calling this out is a key step, I think, toward lessening the power of these variable words. The more they are called out, the more people will be aware of their power, and the words will be made powerless or at least clarified. It is my hope that when these words are used elsewhere in confusing and inaccessible ways, they will no longer be met with compliant silence. Rather, they will be met with the question "What do you mean when you use this word?"

The book entitled *What Is Sustainable Technology? Perceptions, Paradoxes, and Possibilities* invokes Poerksen's theory of plastic words as it performs a similar act of calling out the word "sustainability":

> Does the sheer flexibility of "sustainability" imply that the term is unfit for use? Is it one of those "plastic words" (Poerksen 1995) that need conceptual repair work? In this book we have taken another route by stressing that the conceptual weakness of the term has important consequences: since the co-existence of different articulations allows various interests to be combined, the term allows the alignment of different stakeholders with their own perspectives, so facilitating joint efforts. A single interpretation of "sustainable" would make these joint efforts less likely. The co-existence of different articulations also gives rise to dilemmas and paradoxes. When, for instance, EU guidelines translate the abstract notion of "sustainable use of cars" into a requirement for cars to contain a certain percentage of recyclable material, it forces car design engineers to be more creative and responsible but, at the same time, rules out designs using lighter non-recyclable materials that translate "sustainable use of cars" into "less fuel use." In this book, we have avoided using the terms "sustainability" or "sustainable development" in a narrow prescriptive sense because we wanted to reveal the tensions. For example, calling wind turbines "sustainable" implies that adversaries of wind turbines—such as those arguing in favour of landscape amenity or bird protection—are opposed to sustainable development or are merely addressing minor side-effects of a "good" technology. Moreover, there are futures without wind turbines, without nuclear power stations, without biofuels, and they might all be sustainable. Calling a technology "sustainable" can often be a way of derailing any criticism of it. The concept

of SD [sustainable development] is often criticized as being vague and ill-defined. However, in our view, it is not so much vagueness as levels of analysis that are being confused: at some levels concepts can be defined and calculated.[44]

I don't mean to suggest that this sort of wordy explanation must be done with every instance of a plastic word or jargon. But when an entire argument hinges on one word and that word may be confusing or fall short in any way, that is when it is important to define it. There are certainly other methods and models, tools, and tricks to help scientists write in clearer, more accessible ways. Being aware of the effects of jargon and the importance of clear writing is imperative. Keeping a lay audience in mind is another way to keep from slipping into specialized language without clarification. There is even software called the De-Jargonizer designed to scan documents and highlight jargon terms that might impede comprehension for a non-expert reader. By using this software and adjusting the words to prioritize clarification, for example, scientists may be better able to make hard-to-understand writing more accessible and transparent to the public.

As conservationists, biologists, and invasive species researchers continue to add nuance to their understanding of what constitutes an invasive species (especially under climate crisis), they must also refine their terminology and broaden their metaphors for the general public. By using more precise common language and holistic metaphors, they can help reinvent the current understandings and practices of invasive species ecology for the better. I believe in a reflexive, conversational, social, and communal science. I believe in allowing room and space for public engagement by removing inaccessibility and misleading narratives that can obscure antithetical agendas. Speaking about scientific theories clearly, honestly, and with nuance is fundamental.

The Problem Remains

But none of this gets to the heart of the problem, does it? New conceptual frameworks and more accessible language won't solve mass extinction. If we want owls, if we want forests, if we want a living world, what we really need to do is *live* differently. The truth is, changing how we talk or think about invasive species will only get us so far.

What we need is a paradigm shift, not simply in how we communicate with one another but also in what we undertake—together as a species, with other species, and with the environment at large. In other words, we need to change our relationship with nature and each other.

Capitalism, and its ever-intensifying production and circulation of commodities, is inherently driven to destroy the natural world in pursuit of profit. And it is this same profit, based on human exploitation and ecocidal extraction, that funds the bodies we look to for preserving nature through research and policy. Capitalism fundamentally structures how we relate to nature and each other, obliging us to be combative, competitive, and exploitative with each other and the natural world in order to survive. Therefore, any serious conversation about the preservation of nature must begin by recognizing the necessity of overthrowing the capitalist order and proceed with an exploration of what that might entail.

With the devastation of the climate crisis, we are now rapidly undergoing a mass extinction event. It is estimated that approximately 30 to 50 percent of our planet's species presently alive will be extinct by 2050—which promises an unrecognizable, inhospitable world.[45] For comparison in scope and speed, previous mass extinction events were defined by a loss of around 75 percent of species on a time scale of 2.8 million years.[46] Considering that we've likely already lost about 20 percent of our world's species since 1900, we are on track to lose 50 to 70 percent of earth's species in only 150 years.[47] We are hurtling toward biodiversity extermination in record time.

However, recognizing the state of this climate crisis can provide us a chance to reevaluate the outdated and obsolete models that structure many of our conservation practices. If there is a silver lining, it is that revolutionary social change is often prompted by crisis, and there is no crisis more total than the climate crisis. Standing on the brink of irreversible climate catastrophe forces us to make choices about how we live day to day, with radical implications for our efforts to try to save the world.

How we determine and manage "invasive species" must urgently be reevaluated and contextualized. Current appeals to fight against these species serve as distraction and scapegoating. Our environmental crisis isn't driven by invasive species; it is driven by capitalism. Although some invasive species are extremely harmful and contribute to biodiversity loss—this is not debatable—the conventionally understood narrative and proposed solutions are woefully insufficient to address the true scale and root cause of the issue.

For that, a fundamentally anticapitalist paradigm shift is required so that we might more logically, holistically, ethically, and appropriately consider the best way to preserve our planet.

CHAPTER 5

Solutions

The Good, The Bad, and the Deadly

Capital . . . has a clearly "teleological" character: it has a clear goal—its own expansion—and it pursues this goal relentlessly.

—Endnotes 2

In a finite world with finite resources, a society based on exploitation of people, eternal expansion, and extraction for profit is irredeemable. As Karl Marx writes in the first volume of *Capital*, "Capitalist production, therefore, only develops . . . by simultaneously undermining the original sources of all wealth—the soil and the worker."[1] Capitalism is diametrically opposed to a healthy people and a healthy planet. Therefore, we will not find solutions in capitalism. A society in which everything is oriented around profit cannot solve the problems created by that fundamentally flawed orientation. Nevertheless, it seems a new book offering a quick-fix solution, "green reform," or a Green New Deal to solve the climate crisis is published (and endorsed by Bill Gates) each week without touching the heart of the problem: capitalism.

Ecology Can't Fix Capitalism

Some suggest that we should put our faith in nature to resolve ecological issues on its own. Science writer Fred Pearce's *The New Wild: Why Invasive Species Will Be Nature's Salvation* explores one such solution. The book explores novel ecosystems around the world where native and "invasive" species coexist in a relatively harmonious manner. The conclusions he draws from his research exemplify the stance that by leaving nature alone after anthropogenic interference, we are allowing it to heal naturally. However, though the stories are full of promise, the conclusions Pearce and others draw from this sort of research are incomplete.

One hopeful tale recounts how, at the turn of the fifteenth century, Spanish colonizers began converting the Puerto Rican forests into agricultural landscapes as they prioritized the growing of sugar, coffee, and tobacco. About four hundred years later, the United States took control of the island following the Spanish-American War and expanded the area devoted to farmland, significantly decreasing the native habitat to the extent that only 6 percent of original forest remained by the 1940s.[2] However, shortly thereafter, the market for sugar, tobacco, and other goods grown in Puerto Rico declined, and, where cultivation ceased, these farmlands were reclaimed by forest within an incredibly short time. It is estimated that forest cover increased to 60 percent of the island in the fifteen years between 1959 and 1974. Yet this new forested ecosystem was unlike the pre-colonial forests of Puerto Rico. Over one hundred nonnative tree species introduced by European colonizers had spread across the landscape, much to the dismay of many conservationists who had hoped for the native forest to return. There were even debates about whether ecologists should intervene to destroy new forests as a step in restoring the old ones. But Ariel Lugo, director of the USDA's International Institute of Tropical Forestry in Puerto Rico, showed that these new tree species were not preventing native forests from coming back but rather "paving the way for their return." Pearce quotes Lugo: "The invader trees repaired soils and restored biodiversity. Some provided homes for birds, both native and alien, that subsequently spread the seeds of native plants. With time, many of the more slothful native trees joined the thrusting invaders in the new forests, often now germinated by nonnative insects and birds." These invasive species were restoring the degraded habitat, enabling native species to return, Pearce explains, "helping produce what Lugo calls 'beautifully functioning' new forests, with greater biodiversity than the old forests."[3]

This is a story both of devastating colonization and racism and of native resilience, one that reminds us to reframe our engagement with nature. Pearce also cites a 2009 study that analyzed 240 other scientific studies of degraded ecosystems, tracking what happens without conservationist intervention. "They found 'startling evidence that most ecosystems globally can, given human will, recover from very major perturbations.' More than a third of the ecosystems analyzed showed full recovery, as many again showed partial recovery, and only the remaining 67 showed no recovery [over a span of] about forty to fifty years for forests, where trees have to regrow, and less for other ecosystems."[4] These sorts of "solutions" to biodiversity loss and environments ravaged by conquest and industry invite us to casually accept that the pristine world conservationists fight for is gone and that we should embrace nature's ability to adapt to stressors. Though they ask us to reconsider our approach to invasive species, they insist that we essentially do nothing.

Given the ways some invasive species can help promote biodiversity, fill niches in disturbed ecosystems, and nourish soil and native habitat, it is easy to see how one might conclude that the solution is to leave them to their own devices. While this perspective is a useful counterpoint to the more common idea that every weed is a problem, it is nevertheless grounded in the belief that humans are separate from nature, that nature works best without human interference. But we are part of nature, and we have always interfered, for better or for worse. To leave a habitat alone to fend for itself after human activity has brutalized it is to abandon accountability. At this point, we have a responsibility to work alongside nature to heal the planet.

Pearce argues that "nature often needs aliens and colonists and first-movers."[5] Overemphasizing the happy examples of nature restoring its own biodiversity in the wake of human colonization has structured Pearce as almost an apologist for colonialism. This perspective might restrain conservationists from spraying poison all over an ecosystem, but it also enables nature lovers to stand back and let global capitalism proceed unchecked, its impacts naturalized on the winding road to "biodiversity"—a perspective that would certainly be our downfall. A few pages later, Pearce goes on to defend logging. His reasoning? Logging *only* decreases biodiversity by 25 percent. "Logging?" Pearce asks, "Why not? Toxic waste? Some species will like it. Farming? No worries, when the plows are put away, nature will always come back. But it is not a call to let rip. It simply offers hope and realism."[6] The belief that nature will heal itself if we would only leave it alone emboldens Pearce to excuse the common ecocidal practices of capitalism, justified by a few anecdotes of unexpected positive outcomes. This, of course, will not work.

Embracing disrupted ecosystems and a tendency toward species equilibrium to the extent that we also embrace their reckless devastation? This guarantees the continuation of mass die-offs—of people, plants, and animals—for many years to come. This way of thinking—that we can let capitalism continue while relying on the planet's ability to self-correct—also overlooks the essentially irreversible and unavoidable effects of climate crisis, such as sea level rise, ocean acidification and warming, global temperature increase, and ice sheet melting. These disasters are here, set in motion and locked in. Even if we stop greenhouse gas emissions tomorrow, our capitalist society has already triggered feedback loops that will continue to increase warming for centuries. As frozen bogs and permafrost thaw, for example, methane, a greenhouse gas that is "120 times more powerful than CO_2 at trapping heat and 86 times stronger over a 20-year period," will be released into the atmosphere, accelerating warming at a dizzying pace.[7] As ice in the Arctic melts, the oceans lose that ice's temperature-cooling and reflective qualities, which serve to deflect sunlight and radiation, causing more warming and faster melting of the ice sheets. Meanwhile, about one-fifth of the Amazon rainforests, often called "the lungs of the planet," have already been destroyed thanks to climate crisis and man-made forest fires. If another fifth of the rainforest is lost, scientists warn that it might trigger a catastrophic event known as "dieback," in which the rainforests irreversibly dry out, becoming more prone to fires that then release carbon into the atmosphere, ultimately contributing to even more warming. This phenomenon affects smaller forests across the world as well.[8] Whichever way you look at it, our planet is on track to experience the devastating effects of the climate crisis for a long time. Allowing our ecocidal capitalist society to continue unabated will certainly not result in the miraculous natural resiliency that Pearce and others seem to promote.

Alternate solutions are rooted in permaculture and other regenerative forms of land management. In *Beyond the War on Invasive Species: A Permaculture Approach to Ecosystem Restoration,* permaculture designer Tao Orion argues that invasive species should be studied not just for how they harm ecosystems but also for how they help. Invasives can fix nitrogen, feed pollinators, increase the availability of phosphorus in the soil, clean polluted water, restore watersheds, and serve harmed ecosystems in other important ways. She argues we should approach designing ecosystems in ways that incorporate the beneficial aspects of invasive species without allowing their harmful aspects to run rampant:

> The time has come to acknowledge the ecologically beneficial role that invasive species play in the presence of myriad life-degrading processes and in the absence of well-designed land stewardship models. Species invasions don't just happen to passive, unchanging ecosystems; in some cases, they are the direct result of fluctuations in the larger environment, systemic responses to imbalance or crisis. Invasive species represent a call to engage in creative management that takes these ecosystem services into account. If there is a strong desire to remove or slow the spread of an invasive species, then the ecosystem services that they fulfill must be taken into account as part of the management plan.[9]

Like Pearce, Orion is looking for a silver lining. From her permaculture perspective on ecosystems, she sees environmental distress as an indicator for the lay public to reassert our role in land and ecosystem stewardship. Unlike Pearce, Orion's approach invites human intervention into ecosystem management, which I agree is necessary in restoring biodiversity. Ultimately, however, the solutions Orion offers are merely ways of adapting permaculture into our pre-existing capitalist structure and generally overlook the role of capitalism in perpetuating environmental imbalances. She, too, ultimately ignores the root of the problem.

There are other ways. Indigenous communities care for over a quarter of the world's habitat. By and large, these ecosystems are in better health and have far greater biodiversity than those not tended with Indigenous methods. This distinction comes from the practices and the foci of knowledge that have been passed down for generations among Indigenous communities, including intricate understanding of local ecosystems, cultural and intentional practices focused on minimizing wastefulness, and diversification in the cultivation of species through agroforestry, forest gardens, and controlled burning.[10] These approaches work with the land rather than just letting it be.

Since time immemorial, various Indigenous tribes across North America practiced controlled burning. This act of land stewardship enabled the growth of many different plant species, thereby increasing biodiversity. The practice also helped burn away debris that would allow a wildfire to spread far, wide, and fast, modifying the damage fires would cause. After the colonization of North America, Indigenous people were violently displaced and forbidden to practice (among many things) traditional controlled burning. This has directly led to the scale and ferocity of wildfires that we see today.

As Delilah Freidler writes in "California's Wildfire Policy Totally Backfired. Native Communities Know How to Fix It," the need for controlled burning has been made indisputably clear, and the Karuk tribe of Northern California is practicing traditional controlled burning techniques once again. Though there have been fears that conservationists will co-opt cultural techniques and frustrations that decisions around the logistics of burns are coming from people who are not local nor connected with the land, both the habitat and Karuk tribal members are reaping the benefits of renewing the traditional practice. "'These places are a lot happier when we're here,' said Vikki Preston, a cultural resource technician with the Karuk Tribe who grew up observing burns and has participated in multiple trainings. 'The trees are healthy when we're tending to them, taking really good care of them.' After burns, Karuk schoolchildren take field trips into the forest to gather acorns and materials for basket-weaving, traditional activities made possible by clearing the forest floor." Friedler also interviews Lenya Quinn-Davidson, a fire adviser who, in her work training people to safely adopt pre-scribed burning, "finds inspiration in the Karuk approach to fire. 'We should be striving for the level of connection and personal reflection that Indigenous cultures have with their landscapes,' she said, describing a holistic mindset that non-Natives may need to learn from to care for lands more sustainably. 'We're in an era when we need to find a meaningful place for everyone to work on this, every kind of community member.'"[11] Indigenous philosopher and member of the Citizen Potawatomi Nation Kyle Whyte quotes a Karuk Tribal Member who "views 'climate change as a strategic opportunity not only for Tribes to retain cultural practices and return traditional management practices to the landscape, but for all land managers to remedy inappropriate ecological actions, and for enhanced and successful collaboration in the face of collective survival.'" Whyte explains, "Through rekindling traditional burning connected to many human, plant, fish, and animal interactions, the Karuk climate change strategy renews Karuk knowledge to convene the community members themselves and improve the basis for collaborating with nonindigenous parties."[12] It is this piece that is integral: incorporating a holistic, regenerative, land-based wisdom that is and has been held by many Indigenous cultures in our collective efforts to mitigate climate crisis and support biodiversity and a healthy planet.

But what is the foundation of this Indigenous knowledge? Of course, this answer will differ among tribal members and Indigenous communities across the globe. The Climate Atlas of Canada offers the following definition: "Indigenous knowledge is not uniform across the diversity of Indigenous peoples . . . and cannot be separated from the people who hold it. It is embodied. At the same time, there are some common principles across these knowledge systems which some suggest are important in the context of addressing climate change. For example, the concepts of relationality and stewardship carry teachings that all living things are interconnected and therefore must be respected and cared for."[13] Or, as Whyte explains, "Indigenous knowledges, in the simplest terms, refer to systems of monitoring, recording, communicating, and learning about the relationships among humans, nonhuman plants and animals, and ecosystems that are required for any society to survive and flourish in particular ecosystems which are subject to perturbations of various kinds." Further, "Indigenous knowledges range from how ecological information is encoded in words and grammars of Indigenous languages, to protocols of mentorship of elders and youth, to kin-based and spiritual relationships with plants and animals, to memories of environmental change used to draw lessons about how to adapt to similar changes in the future." He writes, "Indigenous peoples see their knowledges as containing important insights about how to negotiate today's environmental issues; they often see the renewal of their knowledge systems as a significant strategy for achieving successful adaptation planning."[14] There is growing awareness in non-Indigenous communities that Indigenous practices and perspectives are integral to mitigating climate crisis. I would argue that it is imperative for non-Indigenous conservationists to seek guidance from Indigenous land-based knowledge and philosophies—ones embedded in the recognition of the connection between people and nature—that are fundamentally different from capitalist worldviews that perpetuate climate crisis and inform dominant scientific perspectives. It is also imperative that this is done through consent and collaboration, not through appropriation.

Moreover, understanding climate crisis through the teachings, histories, and knowledge of Indigenous cultures reminds us that capitalism and colonialism are fundamental. As Whyte states:

> Colonialism refers to a form of domination in which at least one society seeks to exploit some set of benefits believed to be found in the territory of one or more other societies, from farm land to precious minerals to labor. Exploitation can occur through military invasion, slavery, and settlement. Colonialism often paved the way for the expansion of capitalism, or an economic ideology based on wage-labor that prioritizes growth in monetary profits for the owners of assets as the underlying focus, incentive, and purpose of major human social endeavors. Together, colonialism and capitalism then laid key parts of the groundwork for industrialization

and militarization—or carbon-intensive economics—which produce the drivers of anthropogenic climate change, from massive deforestation for commodity agriculture to petrochemical technologies that burn fossil fuels for energy. . . . Colonially-induced environmental changes altered the ecological conditions that supported Indigenous peoples' cultures, health, economies, and political self-determination.[15]

By naming and understanding colonialism and capitalism as foundational instigators of climate crisis, as Indigenous environmental knowledge helps us do, we can all more easily recognize those false conservationist solutions that hypocritically seek to fight climate crisis while allowing capitalism to continue. Through careful, practiced land stewardship, such reasonable and ecologically beneficial conservation practices work to increase biodiversity, not profit.

Green Solutions Can't Fix Capitalism

Capitalism is fundamentally defined by the accumulation of surplus value and the proliferation of fixed capital—machinery and energy consumption—and this cannot be reformed. The capitalist economy will only continue to increase its demand for energy and for more and more complex "labor-saving" devices because this is how companies stay profitable. Yet we are quickly approaching the point of societal disaster in which capitalism demands MORE! and our planet cries, I HAVE NOTHING LEFT TO GIVE. Capital is, in this sense, locked in a death spiral. When we face the reality of this death spiral head on, we can seek out other solutions. Some of these propose new ways of doing capitalism that prioritize environmentally friendly measures, known as green capitalism or doughnut economics. But these are more quick-fix solutions that allow the root problem to grow and spread.

Green capitalism "is an approach that attempts to use free-market mechanisms to mitigate anthropogenic climate change. Its advocates argue that the market supplies the best means to innovate technological solutions that can compete with existing polluting practices."[16] As such, green capital solutions are inherently deceptive:

they tend to focus on the creation of "green industries" and "green jobs," areas riddled with "green" buzzwords, which sound promising while offering us little more than a drop in the bucket. These industries often refrain from admitting their environmental shortcomings, because that would be bad for business.

Climate activist Greta Thunberg described the false promise of green capitalism in a speech in Milan in 2021. She began by paraphrasing quotes from politicians and world leaders:

"Climate change is not only a threat, it is above all an opportunity to create a healthier, greener and cleaner planet which will benefit all of us." [ironically quoting politicians]

"We must seize this opportunity—we can achieve a win–win in both ecological conservation and high-quality development."

"Fighting climate change calls for innovation, cooperation and willpower to make the changes that the world needs."

"We need to walk the talk—if we do this together, we can do this."

"When I say, 'climate change,' what do you think of? I think of jobs—green jobs—green jobs."

"We must find a smooth transition towards a low-carbon economy."

"There is no planet B" there is no planet blah—blah blah blah, blah blah blah.

"This is not about some expensive politically correct green-assed bunny-hugging or. . ." blah blah blah.

"Build back better" blah blah blah.

"Green economy" blah blah blah.

"Net zero by 2050" blah blah blah.

"Net zero" blah blah blah.

"Climate neutral" blah blah blah.

This is all we hear from our so-called leaders: words—words that sound great, but so far have led to no action.

Our hopes and dreams drown in their empty words and promises.[17]

In a society that keeps increasing its ecocidal tendencies, we are offered sugar-coated "green" explanations that seek to soothe us with promising words. It has never been more important to think critically, to look past the "green" veneer to the reality they obscure. And while green or alternative energy sources are significantly better than toxic coal and gas, they are certainly not benign. Jasper Bernes, in the late *Commune* magazine, explains the not-so-environmentally-friendly core of alternative green energy in his essay "Between the Devil and the Green New Deal": "Nearly every renewable energy source depends upon non-renewable and frequently hard-to-access minerals: solar panels use indium, turbines use neodymium, batteries use lithium, and all require kilotons of steel, tin, silver, and copper." Every alternative energy source requires energy-intensive mining. Solar panels have a very short shelf life; windmills need regular updates. Bernes sums up the fundamental hypocrisy: "Just because a United States encrusted in solar panels releases no greenhouse gases, that doesn't mean its technologies are carbon neutral. It takes energy to get those minerals out of the ground, energy to shape them into batteries and photovoltaic solar panels and giant rotors for windmills, energy to dispose of them when they wear out. Mines are worked, primarily, by gas-burning vehicles. The container ships that cross the world's seas bearing the good freight of renewables burn so much fuel they are responsible for 3 percent of planetary emissions."[18]

Every stage of the process of bringing "green" energy to our communities involves toxic pollution, emission of greenhouse gases, habitat degradation, the use of inhumane labor practices—many of which occur unseen, on the other side of the world, at sites of mining or manufacturing far removed from the product's final destination. These are the horrific side effects of capitalism. Even green or alternative energy solutions are fundamentally at fault so long as they are profit-driven enterprises. On a very practical level, one of the biggest factors in the wastefulness of renewable energy production is the reluctance of these very industries to recycle parts of their products, because it costs less to manufacture new parts than to use recycled material. As explained in the article "Filthy Future: Wind and Solar's Toxic Waste Legacy Problem for Next Generation":

The financial incentive to invest in recycling has never been very strong in solar . . . [and its] production boom has left its recycling infrastructure in the dust. To give you some indication, First Solar is the sole U.S. panel manufacturer we know of with an up-and-running recycling initiative, which only applies to the company's own products at a global capacity of two million panels per year. With the current capacity, it costs an estimated $20–30 to recycle one panel. Sending that same panel to a landfill would cost a mere $1–2. . . . The same problem is looming for other renewable-energy technologies. For example, barring a major increase in processing capability, experts expect that more than 720,000 tons worth of gargantuan wind-turbine blades will end up in U.S. landfills over the next 20 years. According to prevailing estimates, only five percent of electric-vehicle batteries are currently recycled—a lag that automakers are racing to rectify as sales figures for electric cars continue to rise as much as 40% year-on-year. The only essential difference between these green technologies and solar panels is that the latter [double] as a revenue-generating engine for the consumer. Two separate profit-seeking actors—panel producers and the end consumer—thus must be satisfied in order for adoption to occur at scale.[19]

Ultimately, the drive for profit embedded in our society will always hinder truly green solutions. The solutions touted under the banner of green capital tell us that we will ramp up the mining of lithium (for electric car batteries), silicon (for solar panels), and neodymium (for wind turbines) and still continue burning fossil fuels! Green capital and industry do not reduce overall pollution but are themselves products designed to seem green while maintaining a capitalist ethic antithetical to a healthy planet. Although green energy may be a lesser evil, it nevertheless peddles unsustainable and unethical ways of generating energy. Spurred on by capitalism's core drive for growth and profit, rational planning prioritizing any other values (like healthy people and a healthy planet) will be impossible.

Nor are nonprofits an escape from the logic and drive of capitalism. It's not unreasonable to assume that, if the profit motive is the problem at the heart of environmental degradation, then perhaps projects *not* motivated by profit could be the solution. Nonprofits serve as charitable means to address certain needs that society cannot—or will not—otherwise arrange to meet. However, they remain fundamentally unable to address the forces creating these unmet needs in the first place. We can look to food banks for an example. As writer-filmmaker Neil Blakemore argues in his essay "Non-Profits Cannot Liberate Us, but Mutual Aid and Solidarity Can Be a Pathway Forward," food banks are an invaluable asset, collecting excess food that might otherwise be discarded and distributing it to people at no cost. But food banks do not address the issues that lead to food waste and food scarcity; they superficially alleviate the symptoms but do not address the cause. "Moreover," Blakemore argues, "because a food bank relies on capitalist food supply chains, they are simply another part of the chain, not a way to rework it." He adds, "Non-profits also help to contribute to a social understanding that the problem of something like plastic in the ocean is a separate issue from a problem like food waste—even though they are both just symptoms of the environmental degradation capitalist economies specialize in." Nonprofits ultimately serve to undermine and cover up the holistic complexity of social problems. Since the 1970s, states prioritizing profitability have turned away from providing free or low-cost public assistance. Consequently, citizens increasingly depend on NGOs and nonprofits to supply those social programs. But these nonprofits are fundamentally tied to profit. They are a way for wealthy donors and tax-dodging corporate or private family foundations to exert a high degree of control over the shape and character of society. In so doing, they serve to perpetuate a capitalist hierarchy. Blakemore writes,

> Non-profits are deeply tied into the power structures of capitalist systems. That doesn't necessarily make the work they are doing meaningless, but it does mean hoping that they will bring about substantive change is idealistic.

At best they can give a public voice to people who have previously been voiceless. They can educate, and advocate. But once capital gets involved it changes the relationship and can (and does) turn people's trauma or suffering into a commodity for raising funds.

The worst they can do (and big ones often do) is funnel serious energy away from effective means of creating real radical change. In that sense, non-profits are not only not change agents, they are counter-revolutionary counter-insurgent groups that protect capital. This is often true even when the people in them have the "right" politics.[20]

Ultimately, despite being created to aid communities in need, non-profits offer feel-good activities that must always adhere to the funder's parameters, priorities, and ethics. Because they exist and are funded by the capitalist machine, all they can ever offer is a rainbow-colored bandage (perhaps made from recycled materials) for a gushing wound. If they tried to solve the root cause of capitalist inequality, they simply wouldn't be funded.

"Green capital" and green energy are not green, and nonprofits cannot save us. The drive for profit is the problem, and it seeps into every nook and corner of our global society. I have to wonder what our renewable energy sources might look like if the motive for the expansion of this industry were not profit but instead the simple aim of a good life for as many human and nonhuman lifeforms as possible. What other too-expensive-to-consider opportunities to reduce the waste and pollution and ecocidal effects of renewable energy production could we discover and act on? What if our priorities were different? What if cost was not part of the equation? What would it mean to rethink the ways we live, with protecting biodiversity as a principal goal? These are questions we can hardly begin to consider so long as profit is the central aim of our social and economic structure. As French political theorist Gilles Dauvé writes, "Scholarly and 'general public' knowledge stops at what is thinkable in the society that gives rise to and maintains it."[21]

It's not easy. In a world streaked with crises and complexity, it can be a relief to put our faith and funding in the hands of "experts."

But whether elected politicians with ulterior motives, unelected conservationist bureaucrats whose project priorities are based on what will get the most funding, or green energy industries prioritizing price over planet, we should know by now—and from the state of the world—that we have been too long betrayed by the profit-driven systems that influence their actions. There is no funding for, and no capitalist interest in, saving the world. And cost isn't the real obstacle either. As Brett Christophers argues in his book *The Price Is Wrong: Why Capitalism Won't Save the Planet*, it's not that change would be too expensive, it's that change is not profitable enough.

You can't put out a burning house with more fire. We cannot have capital and at the same time a healthy people and planet. The solution will not be found in capital; capital cannot and will not change. A sustainable and non-extractive way of living on and with the planet is fundamentally antithetical to capitalism because capitalism thrives on growth and extraction and profit—which is to say, death on a planet that is finite. So, let's change into a society that tells us that saving the world is the most important thing we could be doing. Let's abandon both despair and false hope—let's stop applauding baby steps and nihilism. Let's work to outlive capitalism—before it kills us. It is time to reorient our solutions, our ideas, our projects, and our work around the question of what change would actually require.

We are composite beings; our bodies simultaneously house ecosystems, like microbiomes of the gut, while we ourselves exist as components of wider systems. There is no end of myself or the beginning of another. Endangered species are extensions of ourselves. They are kin. Solidarity with fellow species is fundamental to our reorienting into a society that sustains rather than destroys the biosphere. But we must put out the fire. Just as invasive species are only judged invasive by their interactions with their environment, so too do we, human beings, only act invasively when we degrade our global ecosystem. Or, as ecology professor Audra Mitchell puts it, "The Anthropocene is not the product of 'humanity' but rather particular segments of it."[22] Our capitalist society perpetuates this environmental destruction, and it is by rejecting that way of living and of organizing our society that we can rekindle a non-invasive way of relating to the earth. Our political and economic system *is* our relationship to the natural world: there is no ecological conservation without abolishing capitalism.

Drawing Inspiration from Movements

Capitalism is what propels nearly every driver of biodiversity loss—climate change, overexploitation, pollution, habitat loss, and the spread of invasive species. It drives the drivers, so to speak. Understanding this, we can extend the metaphor to include the engine, which propels the vehicle. Capitalism is this engine, and it is the power of this engine that is being ignored in "green energy" proposals and many ecology-based solutions. Even as there is recognition that capitalism causes climate crisis, these fields often have a blind spot—they overlook this engine, perhaps willfully at times. To work toward a healthier planet, we must have movements that recognize the necessity of dismantling this engine.

The Stop Cop City movement of Atlanta offers one significant example of people confronting the engine. In this movement, people came together to protect the forest from being demolished and "developed" into a cop training facility and an airport. There was an explicit charge that state and capital are the true enemies of biodiversity. People within the movement recognized that conserving habitat means confronting not just surface-level actors (the developers, the police) but also the ideas and systems of oppression that set them into motion. What other movements similarly involve people coming together to defend nature from capital?

The ZAD

In the 1960s, developers in France sought ways to spread economic growth from Paris to other regions of the country. The idea to build an airport in Nantes was part of that plan but was derailed for several decades as other priorities took precedence. Come the turn of the century, the airport was back on the agenda. Nantes farmers living in the area designated to become an airport were evicted but refused to leave, while activists across France traveled to join them in resistance, building treehouses and barricades to prevent the razing of farms and forestland. According to an article in *Shado Magazine*, "In response to the airport proposal, local residents read out a letter declaring that 'to protect and defend a territory, you need to inhabit it.' They called on groups to support the cause and squat in the empty farms and forests as an act of resistance."[23] This decades-long struggle became internationally known as the ZAD—Zone à Défendre, or the Zone to Defend.

The ZAD comprises just over four thousand acres of mostly traditional farmland and forest. In the late 2000s, the first squatted occupations of evicted farmland in the ZAD marked the beginning of a shift in the resistance, which would evolve into something more akin to a war. In 2012, the French state initiated Operation Caesar, a six-month violent conflict during which they attempted to take control of the ZAD. This drew international attention to the struggle. Police used projectiles, like tear gas and grenades, and stormed the defenders with tanks, helicopters, and drones. Defenders in turn responded with Molotov cocktails, tractors, as well as joyful, festive parties to claim their right to inhabit and militantly guard the land from the vanguard of development. The Zadists built barricades and dug trenches deep into the road to prevent the state from advancing and infiltrating the ZAD. As the police destroyed these blockades, the defenders banded together, rebuilt, and expanded. As described in a 2019 *CrimethInc.* article, "Participants had already established many different kinds of collective infrastructure including bakeries, a pirate radio station, and a legal team; after 2012, these thrived alongside an autonomous healthcare system, large-scale agriculture, and weekly distribution of locally produced food."[24] By 2016, over sixty thousand people were involved in defending the ZAD. In the numerous clashes between protesters and police, about three hundred defenders were injured or hospitalized—a man even had to have his arm amputated. Finally, in 2018, just over a decade after the first squatter occupation of the ZAD, French Prime Minister Édouard Philippe announced that the government would abandon the project to build the airport in Nantes.[25] Following this successful defense of the land, some of those who fought in the struggle have taken up residence at the ZAD. It's now an autonomous zone that is home to around fifty collectives. Residents are practicing a collaborative, horizontalized, and communal way of living, building on the infrastructure created during the struggle:

> This kind of thinking shows us that another world is possible, but it's not going to be a perfect one, and it's not going to be an entirely autonomous utopia. . . . You can never live outside of capitalism: we live in spite of it. As French poet and Surrealist Paul Éluard said: "There is another world, and it is in this one." Another world is possible, and it will rise through the cracks and folds of this one. We will resist this world, and while doing so prefigure a new one alongside it. This absence of a perfect utopia is not a reason to despair. In fact, it's something to celebrate, because that means we can, and have been, building new worlds, as we speak.[26]

The story of the ZAD is arguably a success story that tells us how nature can be protected from the state and development and shows us that another way of living—another world—is possible. While the struggle to defend the ZAD has abated, giving way to an era of tentative calm, members of the ZAD have continued to fight in solidarity with various ecological and social movements across France against the incursion of profit-driven projects that exploit the planet and its people.[27] After all, the ZAD slogan is "Against the airport *and its world*."

Standing Rock

In 2015, the Standing Rock Sioux, whose territories span what is now known as North Dakota and South Dakota, were beginning their own battle against capitalist land developers. Energy Transfer Partners had been granted permission to build the Dakota Access Pipeline (DAPL) by the US Army Corps of Engineers. The 1,172-mile pipeline, which was to be constructed just upstream of the Standing Rock Sioux reservation, would allow the transfer of oil across four states and pass underneath two large rivers. In response, the tribal council "passed a resolution regarding the pipeline stating that 'the Dakota Access Pipeline poses a serious risk to the very survival of our Tribe and . . . would destroy valuable cultural resources.'"[28] But, more than that, the Dakota Access Pipeline and the ecological destruction it would bring represented disrespect to the land and its people that was shocking enough to rally thousands of people to fight against the project.

Originally, the DAPL was meant to cross the Missouri River north of Bismarck, North Dakota, but later plans rerouted the pipeline to pass just upstream of the Standing Rock reservation. The unspoken implication: it is better to risk poisoning the Native people on their reservation than risk poisoning the mostly white residents of the city of Bismarck. This act of environmental racism disregarded both tribal sovereignty and land stewardship, and the threat to the lands of the Standing Rock Sioux caused uproar across the nation and the world. In *Our History Is the Future*, Lakota writer Nick Estes explains the importance of the lands and the river where the pipeline crosses:

> The place where the pipeline crossed the river also held deep historical and cultural significance. Many Horses Heads Bottom, where DAPL crossed the Missouri River, [Standing Rock Elder Phyllis] Young explained, was where Dakotas fled generals Sibley and Sully's 1863 "columns of vengeance." After the 1862 Dakota Uprising, the United States punished survivors of that war at the Whitestone Hill Massacre, where they gunned down more than 400 Lakotas and Dakotas on a buffalo hunt. It was a massacre nearly forgotten by settlers but no less horrific than Sand Creek and Wounded Knee. The soldiers led a manhunt up and down the river, capturing or killing survivors. Mothers plugged their babies' noses to silence their cries as they swam to safety across the river in the cover of darkness.

The Standing Rock pipeline protests began in 2016, when Lakota and fellow members of the Oceti Sakowin peoples established resistance camps close to the construction site. Support for this action grew and intensified, with people coming from across the country to join and hold the camps. While tensions between non-Indigenous participants and Lakota did exist throughout the fight, Estes writes of the connection and collaboration between disparate defenders of Standing Rock, united by a common goal: "Political elites and corporate media have frequently depicted poor whites and poor Natives as irreconcilable enemies, without common ground competing for scarce resources in economically depressed rural areas. Yet, the defense of Native land, water, and treaties brought us together. Although not perfect, Oceti Sakowin camp was a home to many for months. And the bonds were long lasting, despite the horrific histories working against them."[29]

Estes goes on to explain: "The protestors called themselves Water Protectors because they weren't simply against a pipeline; they also stood for something greater: the continuation of life on a planet ravaged by capitalism. This reflected the Lakota and Dakota philosophy of *Mitakuye Oyasin*, meaning 'all my relations' or 'we are all related.'"[30]

These protests turned violent when police attempted to infiltrate and destroy the camps, using dogs, tear gas, rubber bullets, grenades, water cannons, and pepper spray to disperse the Water Protectors. This resulted in hundreds of injuries and many hospitalizations, and the brutality was filmed and shared across the world.[31] Despite nearly a year of struggle at Standing Rock and solidarity movements across the country, the pipeline was finished in 2017, with the reassurance that it would be maintained up to code and all possible safety precautions taken.[32] In 2017, over six thousand barrels spilled from the pipeline, followed by another forty-five hundred barrels in a spill just two years later. Most recently, fourteen hundred barrels spilled in 2022.[33]

Although the pipeline was ultimately built (and inevitably leaked and spilled), the uniting of the Oceti Sakowin and many other Indigenous activists from across the continent at this site was a historic event in the ongoing legacy of Indigenous resistance, moved by the vision and dream of a better future for each other and the planet. As Estes explains:

> Perhaps the answers lie within the kinship relations between
> Indigenous and non-Indigenous and the lands we both

inhabit. There is a capaciousness to Indigenous kinship that goes beyond the human and that fundamentally differs from the heteronuclear family or biological family. "Making kin is to make people into familiars in order to relate," writes Dakota scholar Kim TallBear. "This seems fundamentally different from negotiating relations between those who are seen as different—between 'sovereigns' or 'nations'—especially when one of those nations is a militarized and white supremacist empire." The Water Protectors also ask us: What does water want from us? What does the earth want from us? Mni Wiconi—water is life—exists outside the logic of capitalism. Whereas past revolutionary struggles have strived for the emancipation of labor from capital, we are challenged not just to imagine, but to demand the emancipation of earth from capital. For the earth to live, capitalism must die.[34]

A Growing Ecological Resistance to Capital

Movements like these are among the most sensible responses to the biodiversity problem. Unlike the work of ecologists and "land managers" focused Sisyphean conservation efforts, these popular struggles are examples of movements confronting the core of the issue, the drive for profit. The protests in Atlanta, Standing Rock, and the ZAD are just a few of a growing number of ecological struggles large and small that center the problem of capitalism and, in so doing, serve to truly defend both nature and the people who steward it—who live with, not against, nature.

In the ZAD, the farmland is not industrialized. The locals practice land stewardship in regenerative, careful ways. If you walk along the bocage, the thin strip of forest between farms and fields, you will find ancient chestnut trees that have been coppiced for generations, enabling the people to harvest firewood and lumber while allowing the tree itself to live. There is a sawmill that creates lumber from the forests in the ZAD with which to build homes and infrastructure for the people living there. A mill turns grain into flour, and a cooperative turns that flour into crepes that are then sold as a source of income for the collectives, while farm produce grown or processed through regenerative agricultural practices provide another stream of revenue. Meanwhile, the Lakota have lived in relationship with the land since time immemorial. Upholding the cultural understanding that the land and everything on it are relatives, generations upon generations of Lakota have cared for, nurtured, and developed a deep relationship with the ecology of their unique landscapes, just as those landscapes cared for, nurtured, and developed a relationship with the Lakota. In each case, the connection to the land is obvious and unobscured by capitalist abstraction. As some of the defenders of the ZAD state in an essay titled "Prise de Terre(s)" ("Earth Connection"):

> Our ways of living in and caring for the bocage potentially contain another relationship with nature. A relationship that sees the bocage as a surrounding that envelops and encompasses us, rather than as an exteriority that superior man should manage—whether with the disastrous aim of "exploiting" it or the naïve and presumptuous one of "preserving" it.
>
> *Nous sommes la nature qui se defend* ("We are nature defending itself") is a slogan that has taken the world by storm. It would have been more accurate to say, "We are the bocage defending itself," to remind us that the bocage is not so much "natural" as the fruit of a symbiotic relationship between plants, humans and animals. In fact, even the Amazon rainforest is shaped by a form of gardening that escapes the Western gaze, but not that of its inhabitants.
>
> Extracting ourselves from the artificial binary between man and nature, between the domestic and the wild, enables us to (re)construct a relationship with the world capable of caring for the wild flora and fauna at the very heart of farming territories, as well as for the farming practices at the heart of wild territories. The earth is neither a natural reserve nor an agricultural resource; it's a web of relationships between minerals, plants, animals and humans: a "whole-world" [or "all-world"], in the words of Édouard Glissant.[35]

In order for us to truly begin working toward a healthy planet and a healthy people, we need to reject production for profit and refocus on production for need. We need a revolutionary permanent break from our accumulation-based economic model and obedience to those who uphold it. As Estes explains:

> For the Oceti Sakowin, the affirmation Mni Wiconi, "water is life," relates to Wotakuye, or "being a good relative." Indigenous resistance to the trespass of settlers, pipelines, and dams is part of being a good relative to the water, land, and animals, not to mention the human world. Contrast this with the actions of Energy Transfer Partners (the financial backers of DAPL)—and of capitalism, more broadly, which seeks above all else to extract profits from the land and all forms of life. This is not to suggest that Indigenous societies possess the solution to climate change (and in fact, many Indigenous nations actively participate in resource extraction and capitalist economies in order to strengthen their self-determination). But in its best moments, #NoDAPL showed us a future that becomes possible when everyday Native people take control of their own destinies and lands, while drawing upon their own traditions of resistance.[36]

What do we need? We need nature; we need each other—that is saying the same thing. We need to hear battle cries of "Nous sommes la nature qui se defend!" and "Mni Wiconi!" across the lands and waters. We need to be nature defending itself; we need to be good relatives. These struggles exemplify movements of people who recognize this truth in one way or another: that the time of prioritizing financial growth (and the exploitation of people and planet implied in this pursuit) must end.[37] For people to survive, nature must also be allowed to thrive.

There are no easy solutions, but the principle we must agree on is simple: we must prioritize the health and well-being of our global ecosystems—which includes the people living within them. "Prise de Terre(s)" quotes German philosopher Gustav Landauer: "We can only reclaim the part of nature that belongs to all men if the part of nature that we ourselves are is transformed, if there exists within us a new spirit of renewal of all conditions of life."[38] The way forward is clear. Things need to change. We must bring about a new way of living with the world and with one another.

CHAPTER 6

From the Old World, We Know
A New World Is Possible

*Nothing owns her [Mni Sose, the Missouri River] and therefore she cannot
be sold or alienated like a piece of property. (How do you sell a relative?)*
—Nick Estes

We know what is wrong with our current framework around invasive species management and how it helps distract the public from the ecocidal effects of capitalist society. We know that killing every invasive nonhuman species will not solve these problems and also that changing our framework and language alone will also not solve these problems. So, what will? A fundamental break from capitalism is imperative, but how can we get there, and what could a new future beyond capitalism look like?

The problems of our predominant invasive species story also show us the problems of capitalism. The militaristic metaphor in the "war" against invasive species only exacerbates the belief in the separation between people and nature. Science and science-based policy written in slippery, inaccessible language increases our reliance on a hierarchy of unelected experts to make formal decisions that are, more often than not, motivated by profit and thus fundamentally opposed to the sustainable well-being of people and planet. So, what do we need instead? We need to revitalize a relational way of living with, not against, the earth and to practice a nonhierarchical way of living with one another. This is how we break from capital.

Living with Nature

For most of human history, people have lived as part of nature. As author David Hinton writes in *Wild Mind, Wild Earth*, "Indeed, the concept of 'nature' or 'wild'—everything outside the human realm—would be inconceivable for Paleolithic hunter-gatherers because they had no human cultural space distinct from wild earth, no subjective realm distinct from the objective, and therefore no separation between human and non-human."[1] The understanding that we are of nature—that we *are* nature—is, in many ways, instinctual. In Anishinaabe and Ukrainian author Patty Krawec's book *Becoming Kin*, Krawec speaks of the traditional Anishinaabe term Nii'kinaaganaa, where "Nii" means "'I am' or 'my,'" "Kinaa" means "all of them," and "Ganaa" means "relatives, my relatives." So the term Nii'kinaaganaa "could mean any of these things: I am my relatives, all of them. I am related to everything. All my relations."[2] Nick Estes speaks of a nearly identical concept among the Lakota people—the philosophy of Mitakuye Oyasin, which means "we are all related."[3] Similarly, in the language of the Inuit, Inuk, the word "sila" emphasizes this connection of people to planet. Inuit microbiologist Aviaja Lyberth Hauptmann says, "Sila can mean weather, climate, spirit, wind, consciousness. What that points to is that it's all connected . . . We're not separate from nature; when our minds are well, when the sila inside of us is well, we can treat our environment well."[4] These three terms, traditional to very different land-based peoples of the North Americas, underscore this understanding of an inherent connection with the land.

Historically, it was widely understood that to tend to the land meant to tend to ourselves. Yet there remains this idea that for nature to be natural, it must be untouched by people. A 1995 *Deseret News* article titled "Harmony with Nature? Not in Ancient America" argued, "Contrary to widespread belief, evidence is mounting that pre-Columbian America was not a pristine wilderness inhabited by people who lived in such harmony with nature that they left it unmarked."[5] That the Americas were largely not a wild, unaltered Eden, that Indigenous peoples were influencing and changing the landscape and habitats is certainly true. But this human influence does not make nature unnatural.

As noted, habitats are not passively awaiting the presence of organisms—organisms organize habitats. Consider the beaver. Often called an "ecosystem engineer," the beaver drags logs to build dams, dramatically redirecting the flow of water, creating pools and ponds that significantly increase biodiversity. But even in subtler ways, living things across our planet survive by softly and gradually manipulating, changing, and acting on their environments. Gymnosperms evolved to have attractive flesh-covered seeds so they can be eaten. Creatures that eat the pregnant fruit gather energy and nourishment while the hidden seeds survive their digestive tracts. The seeds are deposited in their droppings where they wander and grow. Millions of years ago, a far-off flower, in need of pollination, grew bright, vivid colors to help attract an ancient, thirsty bee. Beneath our feet, the fungal network spreads nutrients throughout the forest floor, feeding and being fed by plants that thrive through their roots. Every living thing interacts with its environment, changing it to some degree in order to survive. This is not unnatural. This act occurs in perpetuity, and it is these behaviors—this intricate web of relationships—that make up nature, that make nature alive. As we are a part of nature, we too act in these ways. We humans have always gathered fruit, planted seeds, raised animals, tended gardens, and built shelters, but we didn't stop there. Today our way of relating to nature has become a lot more extractive, with less care paid to reciprocity.

I have argued that we are not separate from nature, that people and nature are intertwined, that we come from nature, are a part of nature, and must rekindle a symbiotic relationship with nature. As Marx writes, "Man *lives* on nature—[this] means that nature is his *body*, with which he must remain in continuous interchange if he is not to die. That man's physical and spiritual life is linked to nature means simply that nature is linked to itself, for man is a part of nature."[6] The idea is not new or unique to a given culture, yet today, in a global capitalist society, it is nearly impossible to maintain this ethic. Capitalism "disturbs the metabolic interaction between man and the earth. . . . Thus it destroys at same time the physical health of the urban worker, and the intellectual life of the rural worker," writes Marx.[7] Marx recognized that alienating labor production and the extractive, profit-driven premise of capitalism combine to estrange humanity from our relationship with nature in various ways, depending on our circumstances. Because labor is essentially an involuntary necessity under capitalism, we are alienated from our labor, from the product of our labor (the materials we need to survive), and from the fundamental source of our labor (nature).[8] Marx explains, "Moreover, all progress in capitalist agriculture is a progress in the art, not only of robbing the worker, but of robbing the soil; all progress in increasing the fertility of the soil for a given time is a progress towards ruining the more long-lasting sources of that fertility. . . . Capitalist production, therefore, only develops the techniques and the degree of combination of the social process of production by simultaneously undermining the original sources of all wealth—the soil and the worker."[9] Under capitalism, we are made to plunder nature, extracting and profiting from the environment until there is nothing left. This approach toward the natural world is far more unusual in the scope of human history than our integration within it and marks our collective behavior toward the earth as fundamentally extractive, invasive, and colonizing. As Krawec explains, "Being a settler or a colonizer is not something you *are*; it is something you *do*. It describes your relationship to this land and the people in it. Remember that settlers come to impose a way of living on top of the existing people. Settler colonialism destroys in order to replace. If you are going to stop being a settler and start being kin, that's where we start. With what you do."[10] In this understanding, "settler" describes a way of relating to the world. In an interview with Emmanuel Vaughan-Lee of *Emergence Magazine*, Hinton explains:

> At bottom, what's driving the ecological crisis is our sense of being centers of identity—in the West we call it the soul or the spirit—that are radically separate from, distant from, detached from, what we call nature. And you notice the word "nature," and also the word "wild"—by definition it incorporates these assumptions about identity in the world. And those assumptions are that nature is everything other than us—that is, we're not part of nature and the wild is everything other than us. We're not part of the wild—we're something else. And that I think is what's driving the ecological crisis at its

deepest level, because that distance, that separation, enables a kind of instrumental and exploitative relation to "nature."[11]

Hinton argues that it is this belief of separation that is driving our ecological crisis. But I think this belief is a consequence of our capitalist social system and the ecological crisis it causes, not the source. The capitalist relationship to nature is extractive for profit, nothing more. That separation is a violence forced on us by a capitalist system that demands ever greater profits squeezed from workers who struggle increasingly for their survival.

In our recent collective past, the land belonged to everyone—or perhaps, as the Zadists say, the land belonged to no one. As Estes writes in *Our History Is the Future*, "Mni Sose, the Missouri River, is . . . [a] nonhuman relative who is alive, and who is also of the Mni Oyate, the Water Nation. Nothing owns her, and therefore she cannot be sold or alienated like a piece of property. (How do you sell a relative?)"[12] Under capitalism, however, the land now belongs to some and not others. The land is property, and its owner holds almost absolute say on who can travel on it. This land *can* be sold and used otherwise to make a profit. In industrial agriculture, for example, the landowner pays laborers to work the land, to extract from it products that can turn a profit. The abstraction from land, from labor, and from product embedded in this capitalist order is what brings about this separation between people and the planet. We must change this relationship if we are to save both ourselves and the world.

And we must adapt our society in new ways. We can learn from historical and reciprocal ways of tending to the land, and we must also remember that we cannot go back in time. While we can and must unmake systems of capitalism and colonialism, we will be grappling with their consequences for many generations. Our impact on the earth is essentially irreversible. Increasingly drastic and unpredictable weather patterns, melting ice caps, rising sea levels, mass extinction events, the acidification and warming of our oceans, as well as microplastics that take millennia to fully break down—these are our new realities that we must navigate while building a future that is regenerative and sustaining.[13] This new world—irrevocably altered by climate crisis, capitalism, and colonization—will need new ways of tending to it, but we can look to the past to inform these future practices. Rather than poisoning or bulldozing the earth (or "managing" invasive species) to preserve a dated and pristine ecosystem—a taxidermy landscape—we must work with the land carefully in consideration of the whole habitat and its sustainability. As the defenders of the ZAD say,

> We're far from the conservative ideal of a return to the old forms of farming communities, of simply perpetuating the bocage as it is. If we want to defend the bocage in its integrity, it's certainly not to turn it into an eco-museum! It's a question of leaving room for the experimental, the unprecedented, the unforeseeable, of nurturing its unsuspected future. Inventing a mosaic relationship with the communes, rather than the hegemony of a single form of relationship with the territory. This means leaving room for certain areas to fall fallow, as well as for practical experimentation in agriculture.[14]

Following this philosophy, our relationship with nature could be one of camaraderie—this includes our relationship with what we've come to call invasive species. But what does it look like to apply this way of thinking to invasive species management?

Many Indigenous perspectives recognize that various species and nature itself are kin or family. Thus we can look to the Anishinaabe people for other ways of understanding and "managing" invasive species: "There is no way to translate 'invasive species' into Anishinaabemowin, and the term itself is considered disrespectful." One phrase, *"bakaan ingoji gaa–ondaadag,"* was offered by an Ojibwe Elder to describe those species that the dominant scientific perspectives would refer to as invasive. It can be translated as "'non-local beings,' or those who are from a different place." This term "reflects the Anishinaabe belief that all beings deserve respect and that all were given original instructions by the Creator. Outside their original communities, their actions may not be balanced by the other beings in their new communities, in some cases causing disruption to the normal function and health of an ecosystem."

When this happens, "respectful actions to minimise or prevent the establishment of bakaan ingoji gaa–ondaadag may be required if they pose a threat to local environmental health. Culturally appropriate tactics include not only physical and biological control but also respectful observation, seeking traditional and cultural knowledge from areas where beings may be native, and the creation of new reciprocal relationships through ceremony and harvest."[15] Or, as Anishinaabe researcher Nicholas Reo and anthropologist Laura A. Ogden explain after interviewing members of the Anishnaabe tribe on how to tend to landscapes into which invasive species have moved, "When the purpose of new plants is not obvious, Anishnaabe plant specialists will turn to animals to help them understand how to interact with new species."

Throughout this book, I have argued that invasive species are not inherently good (or bad) but rather misunderstood and overprioritized in conservation tactics. Certainly, there may be a time when a species becomes too robust and we may step in to thin them out, but my hope is that we can do so as stewards of nature. For many of the Anishinaabe interviewed in these papers, relation-based observation is required to understand a new species. Careful, considerate, and ethical removal should be prioritized above all else, and this slower, more meticulous approach to management does not have to be considered a waste of time or resources. After all, what else is more important than respectfully tending to our shared home, this earth, together?

Reo and Ogden quote Anishinaabe elder Kathy LeBlanc: "I'm not so sure about 'invasive species.' I mean a lot of the plants that we use right now are not indigenous to America. They came with the settlers, way before I was born; but they happened to be beneficial health wise to you. The people, over time, learned how to use them."[16] While "some Anishnaabe individuals and tribal governmental programs see introduced species as a significant threat and are combating them using multiple approaches and technologies, including occasional use of herbicides and pesticides," others "repeated [a] caution to not judge plants and animals for attributes beyond their control. As Rita Bulley described, 'I feel bad for those things that are getting introduced, because . . . they don't know they're invasive. They're just growing, doing the only thing they know how to do.'"[17] By and large, interviews revealed that "Anishnaabe tradition bearers are more concerned about an 'invasive land ethic' than the threats of invasive species. Elements of this invasive land ethic include the imposition of Euro-American property ownership regimes, 'command and control' forms of environmental management, and a worldview predicated on the separation of people from nature."[18]

I believe we should work toward a society that does not regard nature as separate but recognizes that it is entangled in us all. Let us abandon the "invasive land ethic" embedded in our militaristic metaphors that haunt the species of this world. We must learn not to regard nature as a source of constant fuel, parasitically sucking the lifeblood from the rivers, the forests, the prairies, the mountains, the seas. Nor should we regard nature as a sterile museum, with guards on duty to prevent anyone from getting too close. I believe we need to create new societies that reaffirm that we are nature—one that is regenerative, restorative, sustainable. This is the only way we are to survive as a species and as a living planet.

Living with Each Other

Importantly, we must also remember that we need each other and are part of one another. Under capitalism, our innate motivation for survival is undergirded by a scarcity mindset that promotes competition and individualism and the hoarding of money above the overall well-being of people and the planet. As much as we must dissolve the separation between people and nature, we must also unravel the harmful capitalist myth that we are all inherently on our own.

In 1928, a speech given by the notoriously racist mine operator and future president Herbert Hoover (who liked to refer to himself as a "self-made millionaire") addressed this American cultural insistence on self-reliance: "We were challenged with a peace-time choice between the American system of rugged individualism and a European philosophy of diametrically opposed doctrines—doctrines of paternalism and state socialism."[19] Decades later during the Cold War, President Ronald Reagan echoed the ideas of rugged individualism as a direct condemnation of communism. Communism implies a common, shared ownership and responsibility for society, and capitalism is essentially its opposite—promoting hierarchies and self-sufficiency.

But human beings are social creatures. As Gilles Dauvé writes in *From Crisis to Communisation*, "Let's understand elementary needs for what they are. Marx was not the first to remark that the main human need is the need for other humans."[20] That is, the necessity of a true community is fundamentally inherent to life and to a life worth living. Despite the capitalist myth of separation and competition, we need each other to survive and to thrive. And there are and have always been other ways to live together.

Evidence suggests that egalitarianism was a common facet of many ancient peoples, particularly among nomadic hunter-gatherers.[21] As an article in *New Scientist* explains:

> For tens of thousands of years, egalitarian hunter-gatherer societies were widespread. And as a large body of anthropological research shows, long before we organised ourselves into hierarchies of wealth, social status and power, these groups rigorously enforced norms that prevented any individual or group from acquiring more status, authority or resources than others. . . . Nor were there strong notions of private property and therefore any need for territorial defence. These social norms affected gender roles as well; women were important producers and relatively empowered, and marriages were typically monogamous. . . . These small-scale, nomadic foraging groups didn't stock up much surplus food, and given the high-risk nature of hunting—the fact that on any given day or week you may come back empty-handed—sharing and cooperation were required to ensure everyone got enough to eat. Anyone who made a bid for higher status or attempted to take more than their share would be ridiculed or ostracised for their audacity. Suppressing our primate ancestors' dominance hierarchies by enforcing these egalitarian norms was a central adaptation of human evolution, argues social anthropologist Christopher Boehm.[22]

Inequality was dangerous for the group, and attempts to act hierarchically (hoarding food, asserting dominance) were understood as fundamentally counterproductive to the well-being of the society. While nomadic hunter-gatherer groups were not all egalitarian societies, my purpose here is to introduce the idea that far more egalitarian societies have in fact historically existed, not limited to time or place. A hierarchical social arrangement is not the sole form. Although we are living in a global capitalist system, there are numerous examples of nonhierarchical ways of living that are being practiced today.

The ZAD is a prime example of an attempt to live communally, with property held in common: "The roughly 300 residents of the zad of Notre-Dame-des-Landes have developed into a plethora of collectives experimenting with different forms of social organization, agricultural and economic practices, as well as relations with their living surroundings. It is organized under the common goal of having consensus-based decision-making shape the development of the zad without state intervention; in other words, it is a place of experimentation with different forms of being human." Though the ZAD consists of several hundred people living in approximately fifty homesteads across four thousand acres, they nevertheless participate in organized communal meetings, share resources and infrastructure (mill, library, radio station, etc.), and are financially supported by the food they grow and sell across the region.[23]

Another prominent example is the Acorn Community Farm in Virginia, which exists as an "egalitarian, income-sharing, secular, anarchist, feminist, and consensus-based commune."[24] The Cambia Community, also in Virginia, is comprised of "cultural experimenters hoping to answer the question of how to build a trust-based and compassionate community that also motivates its members to achieve their highest potential." Notably, Cambia acknowledged the high failure rate of such communal living experiments, stating "given sufficient evidence of failure, we know we must try something different. Please send us your ideas, inspirations, and connect us with others who are doing good work. It's okay, most intentional communities fail. At least 90%."[25]

Some contemporary nomadic cultures demonstrate egalitarian practices that bear similarities to those mentioned above. The Hadza people of Tanzania and the !Kung people of the Kalahari Desert in Southern Africa are two often-cited examples of existing nonhierarchical societies:

> Among the Hadza of Tanzania, the !Kung, and other nomadic hunter-gatherer tribes, resources are shared equally. The concept of ownership, when it exists, generally takes the form of being associated with a place or thing rather than possessing it. Hunted or gathered food is shared equally among all members of the tribe. Sometimes, a chief or elder individual will be responsible for divvying out the food, but this authority figure receives as much as the other tribe members. Generally, individuals who attempt to assert dominance or establish a leadership position are ridiculed and ostracized.[26]

The Hadza people and the !Kung people are unique and complex societies, and it is both inaccurate and disrespectful to draw unnuanced or definitive conclusions about modern nomadic societies and ancient hunter-gatherer cultures. But these examples show that an extractive, hierarchical, and exploitative society is not (and has not been) inevitable. We have evidence of communal living, shared property, and nonhierarchical, non-extractive human relationships throughout history as well as in modern times.

While rare in practice today, the impulse to live more equitably is not unique. In 2020, the World Economic Forum conducted a poll of twenty-one thousand adults from twenty-eight countries, finding that a large majority, 86 percent, expressed the desire for a more egalitarian and sustainable way of living.[27] Regardless of the high failure rate of experimental communities in the modern age, there nevertheless exists an enduring desire for more collective ways of living. It seems that many of us recognize that capitalism is flawed.

Capitalism alienates us from our relationships with nature and one another. But even as capital exploits us and our world, it cannot completely obscure the fact that we need both nature and each other for survival. As Marx explains, "The worker can create nothing without nature, without the sensuous external world. It is the material on which his labor is realized, in which it is active, from which, and by means of which it produces." He continues, "Physically man lives only on these products of nature, whether they appear in the form of food, heating, clothes, a dwelling, etc."[28] It is only through our shared and sustainable manipulation of nature that we are able to provide for each other and ourselves. Socialist writer Judy Cox explains Marx's understanding of our interrelatedness: "Our species being is also a social being, as Marx explained in the *Economic and Philosophical Manuscripts*: 'The individual is the social being.' People have to enter into relationships with each other regardless of their personal preferences because they need to work together to get what they need to live."[29] These social relationships, and our reliance on each other, Marx argues, make up the nature of humanity.

We are social beings, but it is difficult to argue that there is one way for humankind to be, for human nature to manifest. My claim here is not that we are creatures destined to live communally—we need only look at the modern world to see this is not true. But neither are we hard-wired toward selfishness, as some economists, philosophers, and social scientists would claim. Examples of historical and modern practices of egalitarianism—and even failed attempts at and unfulfilled desires for it—show us this.

I argue that what many consider to be "human nature" is changeable, like our natural world, and we are its shapers. Put it in another way, it is in our nature to alter our natural world, and it is in our nature to alter our human behaviors. While inspired by instincts, we are not beholden to them. Although dominance hierarchies may have originated in ancient primate social behavior, interdisciplinary scientist Deborah Rogers points out, we human primates are not stuck with a survival-of-the-fittest social structure. Rogers explains, "We cannot assume that because inequality exists, it is somehow beneficial. Equality—or inequality—is a cultural choice."[30] We are smart animals, and to an almost unbelievable degree we do get to *choose* how to behave. We are also social creatures, and even under an alienating and individualist social structure like capitalism, we rely on one another. At the same time, we are creative and malleable—there is no one way to be human. Marxist scholar Chris Byron explains, "The degree to which people are creative in their labors is an attribute Marx believes is *uniquely* human."[31] We have the unbounded capacity to imagine and to change how we engage with one another and our natural world. We created capitalism, one among thousands of other ways we have invented to live; we can create something else.

So, while I do not argue that human nature is predestined to follow a particular model of society, I can assert that certain societal ethics are in material fact better for people and for the planet. After all, a 2014 study traced societal collapses throughout history (the Egyptians, Romans, and Mayans, among others) to the overexploitation of natural resources or to hierarchical and exploitative economic societies.[32] Such attributes of capitalism are not only undesirable but also likely doomed. If we indeed are social creatures, if we are creative beings, and if we are smart enough to know that a society built on the exploitation of people and planet will likely lead to the demise of ourselves and our habitats, then I believe it is our collective responsibility to take action to change course.

Our Future

We are rapidly approaching a time when every living thing in the world will be dramatically affected by climate crisis. Food and clean water will become scarcer. There will be famine and drought. Certain areas of the planet will become uninhabitable by most living things. There will be mass human and nonhuman migrations across the globe. Climate-related disasters—such as earthquakes, fires, and floods—and extreme temperatures caused by heat domes and polar vortexes will become ever more frequent all over the world. These compounding environmental disasters will wreak havoc, and it is very likely that society as we know it will collapse.

We know we are facing climate crisis, societal crisis, economic crisis, and humanitarian crisis. Crisis is in our future, no matter where we live. When communities are in crises, their best chance at survival is when people come together. Many come to recognize that the capitalist priorities of individual wealth make absolutely no sense in the grand scheme. Mutual aid, collaboration, and collective survival comprise a very natural response to crisis situations. In *A Paradise Built in Hell*, author Rebecca Solnit describes the effect a

1989 earthquake had on residents of San Francisco: "When the subject of the quake came up with a new acquaintance, she too glowed with recollection about how her San Francisco neighborhood had, during the days the power was off, cooked up all its thawing frozen food and held barbecues on the street; how gregarious everyone had been, how people from all walks of life had mixed in candlelit bars that became community centers. . . . Without orders or centralized organization, people had stepped up to meet the needs of the moment, suddenly in charge of their communities and streets."[33]

One often feared worst-case scenario for crisis or disaster is that some people will rise to take advantage of the tumult for their own benefit. The fear is that these people recognize the crisis as an opportunity to take control, seize power, and exert authority, even violently. As Solnit describes, "But to understand both that rising [from ruins of disaster] and what hinders and hides it, there are two other important subjects to consider. One is the behavior of the minority in power, who often act savagely in a disaster. The other is the beliefs and representations of the media, the people who hold up a distorting mirror to us in which it is almost impossible to recognize these paradises and our possibilities."[34] This fear of individuals using the disaster as an opportunity to take power is echoed by the media, she argues, and this false narrative diminishes the real-world capacity of communities to come together in crisis, stoking mistrust and disunity.

Despite what our media would have us believe, most often it is the state that acts violently in these moments. In crisis, the state can become afraid of citizens acting beyond capitalism's priorities and will often exert oppression.[35] For example, after the fires caused by the 1906 earthquake tore through San Francisco, the state was apprehensive that it would no longer be able to control its residents. Solnit writes, "The authorities' fear was not precipitated by anything the public did in those days, but by earlier anxieties in that era of upheaval. They believed uncontrolled crowds routinely degenerated into mobs, and they doubted the legitimacy of the system they dominated, since they expected mobs to tear it apart given the least opportunity. . . . But there is no evidence of civil unrest in the period of the 1906 earthquake."[36]

More than a century after the earthquake, we can recognize how the state's fear of its own subjects continues to this very day and is mobilized when crises, protests, and social movements disrupt the norm of capitalist flow. A 2019 study analyzing decades of international data from the 1950s through 2009 found that the state typically responds to natural disasters by becoming far more repressive. The study showed that the state will decrease democratic conditions by 3.46 percent in the year following the disaster and 10.1 percent over five years, while increasing oppression by 2.5 percent each year.[37] Durham University economics professor Nejat Anbarci says, "It's likely that the disruption that severe storms cause to communities, including the breakdown of day-to-day business and routines, as well as the need for government intervention to support storm recovery, provide an opportunity for governments to tighten their control over citizens, taking advantage of a period of vulnerability."[38] We can expect the growing frequency of climate disasters to present more opportunities for increasing authoritarian control from states around the globe. The state in general appears to believe its citizens will "misbehave" during crises, despite evidence to the contrary. While there may be some people who take moments of chaos to try to assert control and dominate others, this predatory behavior is more reliably the tactic of the state. Individuals and communities, meanwhile, are far more likely to come together to support each other, despite their differences. Solnit writes, "Disaster doesn't sort us out by preferences; it drags us into emergencies that require we act, and act altruistically, bravely, and with initiative in order to survive or save the neighbors, no matter how we vote or what we do for a living."[39]

Perhaps what the state fears at its core is that, through crises, its citizens will realize that capitalism's priorities are both sociopathic and, inasmuch as they are world-destroying, suicidal. Perhaps it fears that people will finally realize that the state and capitalism are the true enemies of our collective well-being, that enough crises will incite people to apply leverage to the cracks they see revealed in our crumbling capitalist society. Perhaps what the state fears is an anticapitalist revolution born of the failure of the state and capital and the collective yearning for a better world.

What We Do Now

How can we bring about an end to capitalism? This is a difficult question to consider yet perhaps the most important one we can ask ourselves and each other. It is also a question with many possible answers. One answer I am fond of is that we can begin planting the seeds of a more desirable world in anticipation of the eventual collapse of our current society. As Neil Blakemore explains,

> A core part of that work is going to be building networks based on solidarity, mutual interest, political education, mass action, defense/safety, worker cooperatives, land trusts, and many more things I'm sure I've never even thought about. In other words, we need to build an alternative social structure within this one. This is called prefiguration or dual power.
>
> Aspects of this process are already underway. Across the country there are growing communities of solidarity economies based in interlocking worker cooperatives, mutual aid, [and] ecological farming principles rooted in indigenous and traditional forms of knowledge. They are nascent, but they are connected and helping each other. There are indigenous leaders who have been protecting the land, water, and environment based on traditional principles.
>
> We need to support these efforts.
>
> Where I live in Los Angeles, there is a dedicated and growing network of mutual aid groups based at least implicitly and often explicitly on the core communist principle—from each according to their ability, to each according to their need. We organize with tenants and unhoused people to get access to food, medicine, harm reduction supplies, clothes, tents, mental health care, medical care, legal aid, eviction defense, court support, protests, and more.

Planting "seeds" of prefiguration or dual power can take many forms—including literal seeds. Grow a garden and share food with your neighbors. Learn how to sustainably forage wild foods and medicines. Join mutual aid and community support groups. Host events—from casual neighborhood block parties to anticapitalist author talks and documentary screenings—to foster social connection and share knowledge. Host skill shares. Take a street medic class. Start a study group. These seeds will propagate an integral network of connections—and planting these seeds of mutual aid and cooperation now will make it easier to rely on each other and survive inevitable crises. As we collaborate in this way, we strengthen the resiliency, autonomy, sovereignty, and connectedness of our communities, which allows us all to benefit from the way crises smooth out differences and clarify priorities. When we come together more frequently, we can help dissolve that myth of rugged individualism that keeps us isolated, separate, and more reliant on capital than on one another. Through these sorts of connections, we can create a web of relationships made up of people with a variety of skills who can be called on in times of need—necessary infrastructure for any society. As Blakemore explains, "By establishing an infrastructure within the current infrastructure we can create the conditions that allow a new society based on the principles of freedom, equality, and solidarity to emerge while the old one collapses."[40] What we can do now is prepare: as more disasters and crises unfold, the institutions of capital may weaken due to instability and widespread disillusionment, just as our mutual aid networks can grow stronger. With broad and skillful networks of autonomous resistance as our base, we can be more likely to push past the limits that have stymied the protest and revolutionary movements of recent years. As capital weakens, alternative ways of living not premised on exploitation and competition can replace it, as long as we take the time now and do the work to strategically build it. At least, this is how the story could go.

It comes back to the stories we are told and the stories we tell ourselves. The story of invasive species is ultimately not an us-versus-them war story, with invasive species on one side and the human protector of habitat on the other side. Invasive species and people are on the same side; the true source of conflict is capitalism. A new ecological, communal, and anticapitalist narrative could make it clear to many across the political spectrum that it is capitalism that is killing us. If more of us take up this story—that we can offer each other a more equitable and sustainable way of living and that it is capitalism destroying both people and planet—then the dangerous social tendency to find and blame scapegoats for collective suffering may hold less sway. As we tell the tale of a better world, we are better able to recognize all the scapegoats that exist to distract us from the true causes of climate crisis and poverty, the sickness of the planet and the sickness of people. But we need to name and define it, we need to be clear in how we tell this story. No more misleading metaphors, no more plastic words. No more scapegoats to distract us. Capitalism is why we and the planet are suffering. When we recognize this, we can more easily band together to fight for a better world—one that will benefit *everyone* and *everything*.

We are living in dark times, and they will get darker. We know the villain of this story—it is capitalism. But how can this story end in hope? The truth is, preserving earth's biodiversity means anticapitalist revolution, wiping away the old hierarchical, accumulation-based political and economic order. It means the reestablishment of communal life, the casting off of all colonial restrictions on Indigenous communities and across the world, and the establishment of global networks for cooperation, including cooperative efforts to mitigate climate change. It means the re-localization of fundamental economic needs (food, water, and medicine). It means radically shortening supply chains, unshackling our communities from bourgeois property law and real estate speculation, and creating regenerative food systems and material-procurement systems unbound from the drive for profit, which can finally center rather than annihilate local and global ecosystems. In short, it means living with one another and with nature once again, and we have the power to bring all this and more into fruition.

Late-stage capitalism is at death's door. What comes next is up to us.

The Colonization of the English Language

A language, like a species, when once extinct, never . . . reappears.
—Charles Darwin

Every creature has a completely different way of understanding the world it inhabits. In her heartrending essay "It Is an Entire World that Has Disappeared," philosopher Vinciane Despret describes the fathomless hole left in the wake of an extinction.

What the world has lost . . . is the unique, sensual, living, warm, musical, and colorful point of view that the Passenger Pigeons created upon it and with it. This unique point of view, to which the world owed the sensation of so many things, is no more. The happiness of being an immense wing traversing infinite spaces; the feeling of being a cloud above earth and of creating changing shapes on it, flowing and shadowy: the sensation of the fields and the woods that, far below, fly by like the images of an accelerating film. The joy of being innumerable and of forming one perfectly attuned being, and the trust in this attunement, which is the figure of joy that the Passenger Pigeons invented when they learned to rely on the air and the wind.

The world has lost the taste of dry and fleshy fruits, of seeds and insects, the raindrops that slide off feathers, the air that dances and that shapes the paths of heat and density, the music in the throbbing murmur of thousands of wings applauding the flight, the creaking of trees and branches shaken under the weight of rest, the shimmer of a rainbow that sweeps in search of the horizon . . . The perception of the vastness, of the innocence [*blancheur*] of an egg, and of the cry of a little one who feels itself abandoned.

All of this is no more. Humanity mourns the Passenger Pigeons. They also say that they should have been concerned, especially when they saw that as they passed in the sky, the sun continued to shine. Humanity can mourn the Passenger Pigeon. But it is the world that bursts with its absence.[1]

It is tragedy absolute, the extinction of a species. In this passage, Despret forces us not to shy away from this heartbreak but to bear witness to this catastrophe as if it were the loss of our own family. To lose the passenger pigeon or the spotted owl or the Galápagos tortoise is to lose a wholly unique way of seeing and being in the world forever. It is losing an entire world, and the aftershock of an extinction can ripple out, causing more endangered species, more extinction. The effect is catastrophic.

With our climate crisis and a capitalist society of consumption, we are experiencing the world's sixth mass extinction. The fate of the passenger pigeon will unfold several million times over, or more. Considering the unfathomable loss of even one species, with its ways of being and seeing the world, the stakes have never been higher, and the cost has never been this dear. We must remember these losses as if they were the loss of a loved one—because they *are* the loss of a loved one. And we must fight with every fiber of our being to not only preserve but also joyfully live alongside the remaining survivors on this earth poisoned by capitalism.

Strengthening ecosystem biodiversity is, as the UN states, the "strongest natural defense against climate change."[2] This will necessitate shifting into a society that prioritizes careful, observation-based land stewardship. It will involve zooming in and out, watching how species interact, allowing ecosystems to move, change, and adapt as our world does the same. This may mean supporting the balance of new, "invasive" species into ecosystems, or it may mean the careful and humane removal of such species. Regardless, I believe we must prioritize not only the biodiversity of our ecosystems but also the diversity of our perspectives and the diversity of voices in our communities.

How is it that we perceive our world?

In 2010, while driving on the highway, I listened to a fascinating episode of the NPR show *Radiolab*, called "Words." The hosts described how the use of words—associating things with names— allows us to perceive the world in a completely different way than if we did not have words. They describe an experiment in which rats are placed in an all-white room with food hidden in one of the corners. The rat sees the food, but, before it can reach it, the experimenters spin the rat or otherwise make it disoriented enough to not remember which corner has the food. What the experimenters found is that the rats, in picking which corner to approach, approached each corner an equal number of times—in other words, it had a 25 percent chance of choosing the correct corner. When the experiment is repeated with one of the walls painted blue and the food always placed in the left-hand corner of the blue wall, the wall can then serve as a "navigational cue" to help the rats understand where the food is. But even with this new blue wall, the rats still only guess the correct corner 25 percent of the time. While rats can recognize blue as a distinct color and left as a distinct direction, they cannot piece these two bits of information together. "Left of the blue wall" is impossible for rats to grasp. In a similar experiment, children up to the age of six will behave as the rats did, likely because their spatial awareness (essentially the idea of prepositions—"of," "under," "on," "through," etc.) takes that long to develop. This experiment suggests that the use of words and the context they provide fundamentally changes how we can *literally* see the world. When one interviewer asked, "What is thought without language?" the other replied, "Well, I don't think it's very much at all."[3] The use of words to name things gives our world shape, depth, and perspective. Those who can speak or understand a language are inheriting with it the ability to see the world in a unique way, fundamentally informed by that culture. Different cultures will carry different contexts—as such, different languages will offer different perspectives and ways of seeing the world. There is a vast system of meaning, interpreting, and perceiving particular to each culture and language. Every language tells a different story.

What happens when one way of seeing and understanding the world dominates? What happens when one language pushes out all the rest?

A 2015 *Atlantic* article titled "The Hidden Bias of Science's Universal Language" asks, "The vast majority of scientific papers today are published in English. What gets lost when other languages get left out?" The author, Adam Huttner-Koros, cites a *Research Trends* study from 2012 that found 80 percent of twenty-one thousand articles indexed in the SCOPUS database were written in English, though they originated from 239 countries.[4] "In some non-English speaking countries, like Germany, France, and Spain, English-language academic papers outnumber publications in the country's own language several times over. In the Netherlands . . . this ratio is an astonishing 40 to 1." In many instances, English scientific jargon, like "'quark' and 'chromosome,' . . . are simply transliterated from English" and adopted, without an equivalent formulated in other languages.[5] Linguist Joe Lo Bianco explains the phenomenon known as domain collapse: "As a language stops adapting to changes in a given field, it can eventually cease to be an effective means of communication in certain contexts altogether." And researcher of science communication Sean Perera states: "The English language plays a dominant role, one could even call it a hegemony. . . . As a consequence, minimal room or no room at all is allowed to communicators of other languages to participate in science in their own voice—they are compelled to translate their ideas into English. . . . In practice, this attitude selects for only a very specific way of looking at the world, one that can make it easy to discount other types of information as nothing more than folklore."[6]

With English dominating scientific communication, scientists from non-Anglophone global cultures are made to relinquish their language in order to be heard. This shunting away of other languages comes at "the great cost of losing their unique ways of communicating ideas."[7] As with the loss or extinction of a species, when we lose a language, we lose an entire worldview. It is this piece that I am primarily concerned with: the space available for other languages and other ways of seeing the world is shrinking.

Historically, Latin was the preferred language of science; today, English is the dominant tongue in which the story of science is told, and English-language dominance extends beyond science. Citing Nigeria and India as reference points, Uwe Poerksen explains that while a diverse array of languages are native to these regions (over 400 in Nigeria and 1,652 in India), English is the predominant language. "Asia and Africa each speak 30 percent of the world's living languages, the Pacific region 20 percent, and the American continent 16 percent. Europe, by contrast, has barely 1 percent—sixty-seven languages. In the birthplace of the nation state, therefore, the number of minorities is the smallest; the number of languages is barely double the number of states. Can we conclude from this that the nation state extinguishes languages? In every case history shows us that the modern state pushes them to the margins."

This global impact on language comes to a head as "a reduction of diversity" and the rise of a monoculture. "Wherever we look," Poerksen notes, "fewer kinds in ever fewer variants of corn and rice and wheat; Chinese, Russian, and English; and sheep, cattle, and pigs look back at us." Poerksen tells us, "Five languages cover almost half the earth, a hundred languages almost all of it. The universalist orientation to the nation state destroys the diversity of living languages. But even these triumphant languages are not the peak of the linguistic pyramid." It is not English itself sitting atop this linguistic pyramid, Poerksen warns us, but "the peak is [composed] of that small and spreading international vocabulary of a hundred, or fifty, or fifteen words."[8] He is speaking, of course, of the tyrannical plastic words that spread to infect languages across borders and cultures.

While diverse languages—over sixty-five hundred—exist across the globe, it is believed that within one or two hundred years the global language count will decrease to just a few hundred.[9] As these languages disappear, "other ways of understanding the world can simply fade away."[10] Today not only is there less room for other languages to exist without pressure, they are also fading away, dying off altogether. They are going extinct.

Wolfgang Sachs compares this extinction of languages to the great loss of biodiversity we are facing as we enter the sixth mass extinction. He writes:

> Languages are dying out every bit as quickly as species. While, in the latter case, plants and animals disappear from the history of nature never to be seen again, with the demise of languages, entire cultures are vanishing from the history of civilization, never to be lived again. For each tongue contains its own way of perceiving man and nature, experiencing joy and sorrow, and finding meaning in the flow of events. To pray or to love, to dream or to reason, evoke different things when done in Farsi, German or Zapotec. Just as certain plants and animals are responsible for the maintenance of large ecosystems, so languages often carry subtle cultures through time. Once species disappear, ecosystems break down; once languages die out, cultures falter.[11]

There is a parallel here between how scientists perceive invasive species and how the English language appears to dominate and colonize as it is swept across the world. As we face the death of other species, we are also facing the death of alternate perspectives, both human and nonhuman. Biodiversity loss and cultural diversity loss march together, stripping the planet.

Grimly, Sachs notes, "Whichever way one looks at it, the homogenization of the world is in full swing. A global monoculture spreads like an oil slick over the entire planet."[12] Industrial civilization seeks to replace the myriad words of the world with one globalized machine language that says "I am." A machine language that says nothing and ostensibly means everything. Wiping out other words, other cultures, making them obsolete, allows only a single narrative to flourish. Shiva writes, "The historical experience of non-western culture suggests that it is the western systems of knowledge which are blind to alternatives."[13] The road to these systems of knowledge is littered with wide-eyed corpses—other ways of seeing. But just as invasive species only act invasively in certain contexts, it is not the English language that is the problem here but instead how it is made to relate to other languages. How it is made to colonize, how it is made to act—this is the problem.

As English is the dominant language of science, plastic words are the dominant words of the powerful, and militarism a dominant metaphor of persuasion, we can see clearly how these factors all combine into a monocultural colonialist perspective in which Western science becomes the dominant knowledge system. Shiva notes, "Dominant scientific knowledge thus breeds a monoculture of the mind by making space for local alternatives disappear, very much like monocultures of introduced plant varieties leading to the displacement and destruction of local diversity. Dominant knowledge also destroys the very conditions for alternatives to exist, very much like the introduction of monocultures destroying the very conditions for diverse species to exist." As Shiva explores in *Monocultures of the Mind: Perspectives on Biodiversity and Biotechnology*, it is an issue of local versus global knowledge. Under the regime of globally dominant way of knowing through science, Shiva argues, "local knowledge is made to disappear by simply not seeing it, by negating its very existence." By local knowledge, she means traditional—often Indigenous—ways of knowing, which are in turn synonymous with (and disparaged as) "primitive" and "unscientific." Shiva writes, "Correspondingly, the western system is assumed to be uniquely 'scientific' and universal. The prefix 'scientific' for the modern systems, and 'unscientific' for the traditional knowledge systems has, however, less to do with knowledge and more to do with power."[14] This dominant perspective would have us equate an Indigenous way of knowing with obsolete belief systems in order to negate and destroy it, winning the ontological competition.

Although "western systems of knowledge have generally been viewed as universal," in this dichotomous clash of local versus global knowledge we are made to forget that the knowledge of capitalism was itself born from a "local system, with its social basis in a particular culture, class, and gender," Shiva says. "It is not universal in an epistemological sense. It is merely the globalized version of a very local and parochial tradition." Our dominant system was once a local perspective that went off to colonize the world, pushing away (often violently) any alternate perspective in its path. Shiva writes, "Modern western science is not to be evaluated, it is merely to be accepted."[15] The scientific worldview is not the only worldview, despite how it is perceived. It is this colonization of belief that needs to be rethought and untaught. It is not that we must wholly abandon Western scientific knowledge. Rather, we must acknowledge that it is one perspective among many and that its roots draw strength (for now) from a capitalist society. The scientific tradition must adopt broader perspectives and abandon its capitalist inclinations.

Dominant scientific knowledge replaces the names of places, plants, and animals with GPS coordinates, nonsensical abbreviations, and dead Latin words. Global capital puts forward terms like "sustainable development" and "scientific progress" as it commits genocide against Indigenous languages as well as their people.

It was out of an elitist desire for hegemony that scientific language was born. Its origins and foundations sought to exclude the commonality, and the commonality is, as it has always been, left vaguely wondering: What is it that is being said here? What is it that is being *done*?

But these are not the questions we should asking, because they do not get to the root of the problem. The questions we should be asking are and have always been: Why is this so? How does this continue to be? And, *what else* can be done?

In the end, this is a problem of culture. Plastic language and militant metaphors are not just tools for the geneticists, the chemists, the ecologists, and the biologists. The plastic tongue works in the mouths of politicians, lawyers, journalists, teachers, students, baristas, retail workers, nannies, and dishwashers. These words are perilous, omnipresent, and often found rolling from your own tongue. Similarly, an us-versus-them understanding between people and nature dominates perspectives.

To argue or converse in these terms and rhetoric is to concede to something you may not consciously agreeing to. When we debate about good or bad *development*, we concede to development across the world. When we argue about how to *manage* invasive species, we concede that they must be managed. And when we speak of battling alien plants and animals that are invading our pristine ecosystems and polluting the purity of native species, we are perpetuating a xenophobic, militant perspective that violently distances us from the natural world.

Plastic words are logical fallacies, tautological linguistic riddles with a lost beginning and no end in sight. They do not invite an answer because they *are* the answer. These words are the ontology of industrial society. This lexicon paints a picture of a world that seeks to carve us into digestible, interpretable data, that wants to police our bodies, our minds, and our spirits. Militant metaphors exacerbate the capitalist ideal of every man for himself and perpetuate antagonistic separation between people and planet. These metaphors pit us against the natural world in a war that will end with no survivors, only death. Together, this language and rhetoric tell a story of the world to be stripped of its natural resources, dominated by human beings and by poison, and project us into a technocratic and capitalist hellscape. In it, we encounter the armored, forward-barreling onslaught commonly referred to as universally desirable economic growth. These words and rhetoric are concepts not to be questioned, and when they so smoothly slip off our tongues, we are made to play our part in a linguistic concession to the ideals of capitalism.

In our work for a better world, the militaristic metaphor is apt—it is us versus them, but our war is against capitalism, not against nature. We are nature defending itself. The world is changing despite capital's attempt to maintain, control, and preserve, so let us let it change for the better. The choice is simple: we can have capital, or we can have life. Let us work toward a human-earth relationship that is not a war.

This book offers one attempt to tear off the veils of the dead machine language, the metaphors, and the stories that render us blind and compliant. These tales we are told about invasive species and their ecological management are taken for granted, as reflecting how the world really is, and then are reproduced in turn. When something is this unquestionable, it inevitably becomes a cultural truth.

This book reminds us that these stories represent a single point of view. Worse, much of what they assert as permanent, natural, and true is false, forged, and temporary. They are stories that use the lexicon and metaphor of industrial civilization to promote its own importance and infallibility.

What can be done?

These stories can be rewritten. We can challenge that story of industrial capitalism; we can challenge its cherished metaphors, prized terms, and seductive scapegoating. We can contest their incontestability and cast them aside. To challenge this story is to begin to rewrite it, and to rewrite it is to step toward a new way of living, without capitalism and consumption and instead with an aim to prioritize care for people and the planet.

We know that simply changing our way of speaking will not change the fate of this world. We must be concerned less with the words themselves and more with how those words are used, how the relationship of those words to other words bolsters the logic of extraction, profit, and deference to power. It is the way our language—the language molded by capitalism—tells us to act that is the problem. Nonnative species are not inherently problematic; they are not harmful *simply because* they are not from here. Similarly, we people living under capitalism are not inherently ecocidal. It is how we are made to relate to our world, its ecosystem and habitats, and each other that we must consider above all else.

David Hinton's perspective is that belief comes first, then action. This refrain is evident in many spaces dedicated to social change: *We must change how we speak and think, and this will inform how we act.* But I believe this is backward: I argue that we must change how we act and relate first. Language, ethics, and society will follow.

This book isn't an answer, but it should serve to remind us that the story of invasive species and the resource-intensive drive to eradicate them is not simply an *environmental, linguistic, or rhetorical crisis* but also a *relational* one.

What can we do?

We can do what Fredric Jameson identified as so difficult for our mindset—we can imagine a new world beyond capitalism, we can tell its story, and we can act now to set that world in motion. The time is right for change. Let us recognize our place in this new world. Let us fertilize the ground and plant seeds to make it grow within and beyond this one. Let us bring this new world to life.

Acknowledgments

I would like to thank many people and some non-people who helped get this book into your hands.

Thanks first to my dear sweet Pete Hamar, for reading this right when we met, which started everything. Thanks for the insight, the edits, and those early conversations that helped enliven not only this writing but eventually all my subsequent writings, all while you were falling in love with me. I love you so much. And thank you for supplying me with the rigorously researched articles, books, and conversations that made the final section make sense.

Thank you to my funny family. Thank you to my dad, Dean Follmann, for your hilariously frank appreciation of my writing and your helpful ponderings. Thank you to my mom, Dona Patrick, for your ceaseless belief in me and everything I do. I owe everything to both of you. Thank you to my brother, Andrew Wyld, for your support, interest, and cheerleading. Thank you to my niece, Lassen, whose very existence reminds me to have hope for the future. I love you guys.

Thank you to my oldest friends. Thank you to my dear friend Bennett Clarkson for consistently working with me all these years and punctuating the exacerbating parts of my work with your clandestine humor and irreverent commentary. Thank you, Nick Nazmi, for the illustrious prank calls. Thank you, Jeb, for your alias.

Thank you to all of the too many editors I ended up recruiting to help me with this. First, thank you, Angelica Sgouros, for your excitement, energy, and patient advice that made this book so much better. I am honored to have you and the AK Press collective believe in me.

Thank you, Steve Kirk, for welcoming with open arms the article this book is based on and publishing it in *Oak*. Thank you for being my first real editor and for your words about my writing. I really think it's what you said that made me feel like this was worth sharing more broadly. Thank you, Clayne Hawley, for your phenomenal advice on the linguistics chapter. Thank you, Miranda Mellis, for ceaselessly working with me when this was a thesis and giving me such consistently incredible feedback. You were so thorough, engaged, and guiding. Thank you, Sara Huntington—you had zero obligation to help with this writing and did so anyway. Thanks for your generosity and phenomenal edits. Thank you, Evergreen State College, for providing a climactic and eventful pair of years for my graduate education, during which this writing was started. Best of luck to you, for as long as you deserve it. And an everlasting thank you to my cherished professors at Sarah Lawrence College, Michael Davis and William "Bill" Shullenberger—without your encouragement, I would never have had faith in my thoughts and ideas. Without your teachings, I never would have learned how to teach myself.

Thank you, mind, books, inspiration, muse, time, life, love, rage, passion, hope, despair, and home. Thanks to all the nonhuman beings on the planet just trying to live. Though it may be a long shot, I hope that the words and ideas in this book help make that more possible.

And thank you to everyone I don't know who ends up reading this. I hope you enjoy it. If you find yourself wanting to talk with me about this book, you can reach me at clarefollmann.com.

Notes

Introduction: The Judas Goats

1. Jesse Hirsch, "Exterminating the Goats of Galápagos," *Modern Farmer*, September 18, 2013, https://modernfarmer.com/2013/09/killing-goats-galapagos.
2. "Project Isabela," *Galápagos Conservancy*, accessed July 18, 2025, https://www.galapagos.org/conservation/project-isabela.
3. Hirsch, "Exterminating the Goats"
4. Daniel Stone, "On the Galápagos, the Betrayal of Judas Goats," *National Geographic*, September 2, 2014, https://www.nationalgeographic.com/culture/article/on-the-galapagos-the-betrayal-of-judas-goats.
5. Hirsch, "Exterminating the Goats."
6. Maricruz Jaramillo, "What Do Galápagos Hawks Eat After the Goats Are Gone?" *British Ornithologists' Union*, July 11, 2022, https://bou.org.uk/blog-jaramillo-galapagos-hawks.
7. Galápagos Conservancy, "Project Pinzón" June 9, 2022, https://www.galapagos.org/conservation/project-pinzon.
8. Rachel Pannett, "Galápagos Giant Tortoises Being Killed for Meat, Officials Fear," *Washington Post*, August 31, 2022.
9. Phoebe Weston, "Lemurs and Giant Tortoises Among Species at Risk If Global Heating Hits 3C," *Guardian*, April 9, 2021.
10. Climate refugees are migrating species whose preferred habitat has been made inhospitable by way of the climate crisis and are therefore forced to flee their habitats and homes. Climate refugees can be animals, plants, or people—any living species whose home has become uninhabitable or dangerous to live in because of climate crisis.
11. "Project Isabela," Galápagos Conservancy.
12. Throughout this book, when I speak of science, I am referring primarily to a dominant Western science. I explore aspects of non-Western science in later chapters.
13. I use the term "anthropogenic climate crisis" to emphasize that climate crisis results from the impact of certain human activity on the planet. However, many people are only contributing to the climate

165

crisis because they live under a larger capitalist system. As Indigenous philosopher Kyle Whyte (tribal member of the Citizen Potawatomi Nation) explains, "'Anthropogenic climate change' or 'the Anthropocene,' then, are not precise enough terms for many Indigenous peoples, because they sound like all humans are implicated in and affected by colonialism, capitalism and industrialization in the same ways. . . . The Anthropocene is rooted in colonization. For colonialism has always included terraforming that tears apart what they call . . . the 'flesh' of human–nonhuman-ecological relationships." An involuntary reliance on a global capitalist society, not its people, is to blame for climate crisis. See Kyle Whyte, "Indigenous Climate Change Studies: Indigenizing Futures, Decolonizing the Anthropocene," *English Language Notes* 55, no. 1–2 (March 1, 2017): 153–62, https://doi.org/10.1215/00138282-55.1-2.153.

14. In *Weekend at Bernie's*, two employees find their boss dead in his mansion before a scheduled party. To avoid suspicion and enjoy their dead boss's home for the weekend, they dress him in sunglasses to convince guests that he is alive. Similarly, despite ample scientific evidence to suggest that our current understanding of invasive species is obsolete, this outdated war story remains the popularized way of thinking about them.

15. As one who is tender-hearted toward the theory of communism, I want to point out that how nation-states have practiced communism is not reflective of how communism must (or should) be practiced.

Chapter 1: The Nature of Language

1. *Oxford English Dictionary*, s.v., "semiotics (n.)," accessed April 11, 2018, https://www.oed.com/dictionary/semiotics_n#.

2. Thomas A. Sebeok, *Signs* (Toronto: University of Toronto Press, 2001).

3. K. M. Newton, "Roland Barthes: 'Science Versus Literature,'" in *Twentieth-Century Literary Theory* (London: Palgrave, 1997), 97.

4. Michel Foucault, *This Is Not a Pipe*, trans. James Harkness (Berkeley: University of California Press, 1982), 5.

5. Words (symbol signifiers) on their own are totally powerless. They are empty shells, ineffective until they are filled with deeper symbolic interactions and, perhaps more importantly, collective agreement that the words mean something. The power of words is fundamentally illusory and only comes from general consensus—a lesson well worth remembering in fields beyond invasive species.

6. *Oxford English Dictionary*, s.v., "metaphor, (n.)," accessed April 11, 2018, https://www.oed.com/dictionary/metaphor_n#.

7. Colby Phillips, "The Effects of Metaphors," *Pen and the Pad*, accessed January 22, 2018, https://penandthepad.com/effects-metaphors -8758991.html.

8. William Shakespeare, *The Tragedy of Romeo and Juliet* (New Haven: Yale University Press, 1954), 36.

9. Phillips, "The Effects of Metaphors."

10. Colin Koopman, "The Algorithm and the Watchtower," *New Inquiry*, 2015, https://thenewinquiry.com/the-algorithm-and-the -watchtower.

11. Eula Biss, *On Immunity: An Inoculation* (Minneapolis: Graywolf Press, 2014), 56.

12. George Lakoff and Mark Johnson, *Metaphors We Live By* (Chicago: University of Chicago Press, 2011), 4.

13. Alan Gross, *The Rhetoric of Science* (Cambridge, MA: Harvard University Press, 1996), 80.

14. Philip Ball, "A Metaphor Too Far," *Nature*, February 23, 2011, https:// doi.org/10.1038/news.2011.115.

15. Richard Lewontin, "The Price of Metaphor," *In Ideas on the Nature of Science*, ed. David Cayley (Fredericton, NB: Goose Lane Editions, 2009), 278–79.

16. Lewontin, "The Price of Metaphor," 279.

17. Susan Sontag, *Illness as Metaphor; and AIDS and Its Metaphors* (New York: Doubleday, 1990), 99.

18. Franco Berardi, *The Uprising: On Poetry and Finance* (Los Angeles: Semiotext(e), 2012), 157.

19. Fredric Jameson, *The Seeds of Time* (New York: Columbia University Press, 1994), xii.

20. It is worth mentioning that "resource" and "product" are both plastic words.

21. Lewontin, "The Price of Metaphor," 278.

22. Lewontin, "The Price of Metaphor," 277.

23. Aristotle, *Rhetoric*, trans. W. Rhys Roberts (Dover, 2004), 123.

24. Shakespeare, *The Tragedy of Romeo and Juliet*, 37.

25. Uwe Poerksen, *Plastic Words: The Tyranny of a Modular Language*, trans. Jutta Mason and David Cayley (University Park: Pennsylvania State University Press, 2004).

26. Poerksen, *Plastic Words*, 1.

27. Poerksen, *Plastic Words*, 62.

28. Poerksen, *Plastic Words*, 15.

29. Ludwig Wittgenstein and G. E. M. Anscombe, *Philosophical Investigations* (New Jersey: Prentice Hall, 1958), 14.

30. Poerksen, *Plastic Words*, 22–23.

31. Mustapha Khayati, "Captive Words: Preface to a Situationist Dictionary," Anarchist Library, 1966, https://theanarchistlibrary.org/library/mustapha-khayati-captive-words-preface-to-a-situationist-dictionary.
32. Poerksen, *Plastic Words*, 94.
33. Poerksen, *Plastic Words*, 85.
34. See, for another example, the seventeen variants of the word "normal" in the brief overview written by Gabriela Beltre and Magda G. Mendez, "Child Development," *StatPearls*, accessed July 19, 2025, https://www.ncbi.nlm.nih.gov/books/NBK564386.
35. *Oxford English Dictionary*, s.v., "develop, (v.)," accessed August 3, 2018, https://www.oed.com/dictionary/develop_v.
36. Wolfgang Sachs, *Planet Dialectics* (New York: Zed Books, 1999), 3.
37. Catherine Dale, *Operation Iraqi Freedom: Strategies, Approaches, Results, and Issues for Congress* Congressional Research Service, 2009, https://fas.org/sgp/crs/natsec/RL34387.pdf. Italics mine.
38. "Iraq Body Count," accessed September 9, 2025, https://www.iraqbodycount.org/database.
39. Poerksen, *Plastic Words*, 7.
40. Michel Foucault, *The Order of Things: An Archaeology of the Human Sciences* (London: Routledge, 2004), xix.
41. Poerksen, *Plastic Words*.
42. Poerksen, *Plastic Words*, 76.
43. Biss, *On Immunity*, 133.
44. Anne Carson, *Eros the Bittersweet* (Princeton, NJ: Princeton University, 2016), 73.
45. Poerksen, *Plastic Words*, 23.
46. Poerksen, *Plastic Words*, 1.
47. *Oxford English Dictionary*, s.v., "manage, (n.)," accessed April 11, 2018, https://www.oed.com/dictionary/manage_n.
48. "Executive Order 13751—Safeguarding the Nation From the Impacts of Invasive Species," Administration of Barack Obama, November 5, 2016, https://www.govinfo.gov/content/pkg/DCPD-201600818/pdf/DCPD-201600818.pdf.
49. Federal Trade Commission, *Strategic Plan for Fiscal Years 2018 to 2022*, accessed July 19, 2025, https://www.ftc.gov/system/files/documents/reports/2018-2022-strategic-plan/ftc_fy18-22_strategic_plan.pdf.

Chapter 2: Science's Complicated Courtship with Language

1. *Oxford English Dictionary*, s.v., "culture, (n.)," accessed April 24, 2018, https://www.oed.com/dictionary/culture_n.
2. These are examples of laws of nature that are considered literally and

objectively true and so must be in accordance with the following five principles: "1. Are factual truths, not logical ones; 2. Are true for every time and every place in the universe; 3. Contain no proper names; 4. Are universal or statistical claims; and 5. Are conditional claims, not categorical ones" ("Laws of Nature" Internet Encyclopedia of Philosophy, https://iep.utm.edu/lawofnat/#H1, access date September 5, 2025). While it is true that there has been discussion around whether laws of nature could change or have changed in the past, that is not the subject of this chapter. This chapter explores issues that can arise when policies are made based on certain scientific theories or in the name of scientific understandings or on beliefs informed by subjective culture, not the laws of nature.

3. "Skulls in Print: Scientific Racism in the Transatlantic World," University of Cambridge, March 19, 2014, https://www.cam.ac.uk/research/news/skulls-in-print-scientific-racism-in-the-transatlantic-world.

4. Donna Haraway, *Simians, Cyborgs, and Women: The Reinvention of Nature*, New York: Routledge, 1991), 8.

5. Charles Darwin, *The Descent of Man and Selection in Relation to Sex* (London: Murray, 1889), 362.

6. Darwin, *The Descent of Man and Selection in Relation to Sex*, 356.

7. Carolyn Merchant, "The Scientific Revolution and the Death of Nature," *Isis* 97, no. 3 (2006): 521.

8. Merchant, "The Scientific Revolution," 519.

9. Merchant, "The Scientific Revolution," 519.

10. Clifford D. Conner, *A People's History of Science: Miners, Midwives, and "Low Mechanicks"* (New York: Nation Books, 2005), 369.

11. Vandana Shiva, *Monocultures of the Mind: Perspectives on Biodiversity and Biotechnology* (New York: Zed Books, 1993), 10.

12. John K. Gilbert and Susan Stocklmayer, eds., *Communication and Engagement with Science and Technology: Issues and Dilemmas* (New York: Routledge, 2012), viii–x; Lorraine Daston and P. Galison, "The Image of Objectivity," *Representations*, no. 40 (1992): 81–128.

13. Daston and Galison, "The Image of Objectivity."

14. Frances Bacon, *New Atlantis, and The Great Instauration* (West Sussex, UK: Harlan Davidson, 1989), 71.

15. "Skulls in Print."

16. Conners, *A People's History of Science*, 358, 359.

17. Conners, *A People's History of Science*, 359; Gilbert, "Communication and Engagement with Science and Technology."

18. Tom Feilden, "Most Scientists 'Can't Replicate Studies by Their Peers,'" BBC News, February 22, 2017, https://www.bbc.com/news/science-environment-39054778.

19. Conner, *A People's History of Science*, 306.
20. Conner, *A People's History of Science*, 353.
21. "The Protestant Reformation," National Geographic Education, accessed July 20, 2025, https://education.nationalgeographic.org/resource/protestant-reformation.
22. Jeremy J. Priest, "Forbidden Translations? A Brief History of How the Mass Came to Be Rendered in the Vernacular," *Adoremus*, October 10, 2020, https://adoremus.org/2019/01/forbidden-translations-a-brief-history-of-how-the-mass-came-to-be-rendered-in-the-vernacular.
23. Conner, *A People's History of Science*, 362.
24. Daston and Galison, "The Image of Objectivity," 81.
25. K. M. Newton, "Roland Barthes: 'Science Versus Literature,'" in *Twentieth-Century Literary Theory* (London: Palgrave, 1997), 94.
26. Daston and Galison, "The Image of Objectivity," 84.
27. Jorge Luis Borges, *Collected Fictions*, trans. Andrew Hurley (Penguin, 1999), 325.
28. Wolfgang Sachs, *Planet Dialectics* (New York: Zed Books, 1999), 44.
29. Daston and Galison, "The Image of Objectivity," 93.
30. Manuel Velasquez, *Philosophy* (Boston: Wadsworth/Cengage Learning, 2014).
31. Richard Stivers, *Technology as Magic* (New York: Continuum, 2006).
32. Poerksen, *Plastic Words*, 43.
33. Hannah Arendt, *The Human Condition: Second Edition* (University of Chicago Press, 1998), 3–4.
34. *Oxford English Dictionary*, s.v., "jargon, (*n.*)," accessed April 11, 2018, https://www.oed.com/dictionary/jargon_n1.
35. Jordana Rosenberg, "Scientific Jargon," Thompson Writing Program, Duke University, 2012, https://twp.duke.edu/sites/twp.duke.edu/files/file-attachments/scientific-jargon.original.pdf.
36. Marcella Hu and Paul Nation, "Unknown Vocabulary Density and Reading Comprehension," ResearchGate, January 1, 2000, https://www.researchgate.net/publication/234651421_Unknown_Vocabulary_Density_and_Reading_Comprehension.
37. Ralf Barkemeyer, "Jargon-Free Science," *Chemistry and Industry* 80, no. 1 (2016): 20.
38. Silja Samerski, "Risk-Anxiety and the Myth of Informed Decision Making," *Thinking After Illich*, https://pudel.samerski.de/pdf/SamerskiOslokorr.pdf, 3.
39. Samerski, "Risk-Anxiety," 6.
40. Tzipora Rakedzon, Elad Segev, Noam Chapnik, Roy Yosef, and Ayelet Baram-Tsabari, "Automatic Jargon Identifier for

Scientists Engaging with the Public and Science Communication Educators," *PLOS ONE* 12, no. 8 (2017), https://doi.org/10.1371/journal.pone.0181742; Aviv J. Sharon and Ayelet Baram-Tsabari, "Measuring Mumbo Jumbo: A Preliminary Quantification of the Use of Jargon in Science Communication," *Public Understanding of Science* 23, no. 5 (2014): 528–46.

41. Rakedzon et al., "Automatic Jargon Identifier for Scientists"; Sharon and Baram-Tsabari, "Measuring Mumbo Jumbo."
42. Barkemeyer, "Jargon-Free Science."
43. K. Horowitz and D. Walter-Toews, letters to the editor, *New Yorker*, March 26, 2018.
44. D. Lewis, letter to the editor, *New Yorker*, March 26, 2018.
45. Amanda MacMillan, "Global Warming 101," Natural Resources Defense Council, December 20, 2024, https://www.nrdc.org/stories/global-warming-101. Not all anti-vaxxers were Trump followers. Many maintain criticism of Western science and medicine given its sordid racist or misogynistic history, but this widespread mistrust of Western science is a clear indicator that we desperately need a new way of thinking about and talking about science—one that is anti-oppressive and anticapitalist. Lisa Friedman and Brad Plumer, "Trump's Response to Virus Reflects a Long Disregard for Science," *New York Times*, April 28, 2020.
46. Erick H. Turner et al., "Selective Publication of Antidepressant Trials and Its Influence on Apparent Efficacy: Updated Comparisons and Meta-Analyses of Newer Versus Older Trials," *PLOS Medicine* 19, no. 1 (2022), https://doi.org/10.1371/journal.pmed.1003886.
47. Turner et al., "Selective Publication of Antidepressant Trials."
48. Johann Hari, *Lost Connections: Uncovering the Real Causes of Depression—and the Unexpected Solutions* (New York: Bloomsbury, 2018), 31.
49. Dalmeet Singh Chawla, "1 in 7 Scientific Papers Is Fake, Suggests Study That Author Calls 'Wildly Nonsystematic,'" *Retraction Watch*, September 24, 2024, https://retractionwatch.com/2024/09/24/1-in-7-scientific-papers-is-fake-suggests-study-that-author-calls-wildly-nonsystematic.
50. Kelly Crowe, "When Scientists Want Their Data Fudged and Why You Should Care," CBC, October 13, 2018, https://www.cbc.ca/news/health/second-opinion-scientists-data-fudging-1.4861556.
51. Donald Trump, "Statement by President Trump on the Paris Climate Accord," White House, June 1, 2017, https://www.whitehouse.gov/briefings-statements/statement-president-trump-paris-climate-accord. Italics mine. Note that "production" and "development" are

plastic words. For a full list of Uwe Poerksen's plastic words, see Chapter 1.
52. Poerksen, *Plastic Words*, 75.

Chapter 3: The Rhetorically Charged, Ethically Dubious, Xenophobic, and Capitalistically Motivated Story of Invasive Species Management

1. In war, broad retaliation against a single "aggressor" is called collective punishment and is a war crime. Here we have a war crime against the natural world under the banner of conservationism.
2. Mark A. Davis, *Invasion Biology* (Oxford: Oxford University Press, 2009), 5.
3. Davis, *Invasion Biology*, 5.
4. Mark Davis, "Researching Invasive Species 50 Years After Elton: A Cautionary Tale," in *Fifty Years of Invasion Ecology: The Legacy of Charles Elton*, ed. David Richardson (Chichester, UK: Wiley-Blackwell, 2011), 40.
5. Davis, *Invasion Biology*, 6.
6. Davis, "Researching Invasive Species 50 Years After Elton."
7. Davis, *Invasion Biology*, 7.
8. Gloria Anzaldúa, *Borderlands/La Frontera*. (San Francisco: Aunt Lute Books, 1987), 3.
9. Daniel Simberloff, "Confronting Introduced Species: A Form of Xenophobia?," *Biological Invasions* 5, no. 3 (January 1, 2003): 181, http://fwf.ag.utk.edu/mgray/wfs560/Biological_Invasions.pdf. Brackets in original.
10. Simberloff, "Confronting Introduced Species," 183.
11. Jens Jensen quoted in Simberloff, "Confronting Introduced Species," 183.
12. Lorraine Daston and P. Galison, *Objectivity* (Brooklyn: Zone Books, 2007), 338. The word "flora" is being used here to refer to people.
13. Charles Elton, *The Ecology of Invasions by Animals and Plants* (Chicago: University of Chicago Press, 2000), 15. Italics mine.
14. Jing Hua Chiu et al., "Trends in the Direction of Global Plant Invasion Biology Research over the Past Two Decades," *Ecology and Evolution* 13, no. 1 (January 1, 2023), https://doi.org/10.1002/ece3.9690.
15. H. P. Lovecraft, *Supernatural Horror in Literature* (Auckland: Floating Press, 1973), 12. It is worth noting that Lovecraft himself was deeply racist, and his prejudice and xenophobia are played out in many of his horror stories.
16. Davis, *Invasion Biology*.
17. Wendell Berry, *Home Economics: Fourteen Essays* (New York: North Point Press, 1994), 18.

18. Brendon Larson, "The War of the Roses: Demilitarizing Invasion Biology," *Frontiers in Ecology and the Environment* 3, no. 9 (2005): 495, https://escholarship.org/content/qt8366cowq/qt8366cowq.pdf.

19. Haila quoted in Brendon Larson, "Who's Invading What? Systems Thinking About Invasive Species," *Canadian Journal of Plant Science* 87, no. 5 (2007): 993–99, http://dx.doi.org/10.4141/cjps071161.

20. Larson, "Who's Invading What?," 995.

21. After all, another word for "invasive species" is "alien."

22. Larson, "Who's Invading What?," 995.

23. Corinne Boyer, *Under the Bramble Arch* (Woodbury, MN: Llewellyn, 2020), 93–94.

24. "Scotch Broom Identification and Control—King County, Washington," accessed September 9, 2025, https://kingcounty.gov/en/dept/dnrp/nature-recreation/environment-ecology-conservation/noxious-weeds/identification-control/scotch-broom.

25. Sebastián A. Ballari, Sara E. Kuebbing, and Martin A. Nuñez, "Potential Problems of Removing One Invasive Species at a Time: A Meta-Analysis of the Interactions Between Invasive Vertebrates and Unexpected Effects of Removal Programs," ed. M. Sánchez, *Peer J Life and Environment* 4 (2016), https://doi.org/10.7717/peerj.2029.

26. Evan C. Fricke et al., "The Effects of Defaunation on Plants' Capacity to Track Climate Change," *Science* 375 (2022): 210–14, https://doi.org/10.1126/science.abk3510.

27. Quoted in Isobel Whitcomb, "Invasive Species Aren't Always the Bad Guys," *Sierra*, February 17, 2022, https://www.sierraclub.org/sierra/invasive-species-aren-t-always-bad-guys.

28. Adolf Hitler, *Mein Kampf* (New York: Reynal and Hitchcock, 1941), 396, 606.

29. Tom Norton, "Fact Check: Is Donald Trump, Adolf Hitler Viral Quote Comparison Accurate?," *Newsweek*, November 15, 2023, https://www.newsweek.com/fact-check-donald-trump-adolf-hitler-viral-quote-comparison-accurate-1843501.

30. Bethany Augliere, "Effort to Prevent 'Coywolf' Hybrids Is Working, Study Finds," *Science*, April 14, 2016, https://www.science.org/content/article/effort-prevent-coywolf-hybrids-working-study-finds.

31. It is ironic that so many conservation efforts attempt to preserve the pure blood and genetics of these wolves while elsewhere wolves are systematically hunted due to the misconception that they cause extensive livestock loss. Cultural predispositions coupled with economic motivation spell death for gray wolves while red wolves are placed on a pedestal. See Ted Williams, "America's New War on Wolves and Why It Must Be Stopped," Yale E360, February 17,

2022, https://e360.yale.edu/features/americas-new-war-on-wolves-and-why-it-must-be-stopped.

32. Robert Colautti and Hugh MacIsaac, "A Neutral Terminology to Define 'Invasive' Species," *Diversity and Distributions* 10, no. 2 (2004): 135–41.

33. Davis, *Invasion Biology*.

34. David M. Richardson, *Fifty Years of Invasion Ecology: The Legacy of Charles Elton* (New York: John Wiley and Sons, 2011).

35. Poerksen, *Plastic Words*, 1.

36. Poerksen, *Plastic Words*, 26.

37. Poerksen, *Plastic Words*, 59.

38. "Executive Order 13751—Safeguarding the Nation from the Impacts of Invasive Species," administration of Barack Obama, November 5, 2016, https://www.govinfo.gov/content/pkg/DCPD-201600818/pdf/DCPD-201600818.pdf.

39. M. Chew and Andrew Hamilton, "The Rise and Fall of Biotic Nativeness: A Historical Perspective," in *Fifty Years of Invasion Ecology: The Legacy of Charles Elton*, ed. David Richardson (Chichester, UK: Wiley-Blackwell, 2011), 36–45.

40. Richard C. Lewontin, *Biology as Ideology: The Doctrine of DNA* (New York: Harper Collins, 1993), 109.

41. "Could Be First-Ever Sighting of Common Dolphins in the Salish Sea," *Journal of the San Juan Islands*, June 14, 2016, www.sanjuanjournal.com/news/could-be-first-ever-sighting-of-common-dolphins-in-the-salish-sea.

42. Eric Holthaus, *The Future Earth: A Radical Vision for What's Possible in the Age of Warming* (New York: HarperCollins Publishers, 2020), 26–27.

43. Mark Davis, "Invasive Plants and Animal Species: Threats to Ecosystem Services," in *Climate Vulnerability: Understanding and Addressing Threats to Essential Resources*, ed. Roger A. Pielke (Amsterdam: Academic Press, 2013), 51–59.

44. Thor Hanson, *Hurricane Lizards and Plastic Squid: How the Natural World Is Adapting to Climate Change* (London: Icon Books, 2023).

45. Marina Bolotnikova, "It's Time to Stop Demonizing 'Invasive' Species," *Vox*, November 28, 2021, www.vox.com/down-to-earth/22796160/invasive-species-climate-change-range-shifting.

46. Davis, "Invasive Plants and Animals," 51.

47. Chew and Hamilton, "The Rise and Fall of Biotic Nativeness," 40.

48. Simberloff, "Confronting Introduced Species."

49. National Invasive Species Council, *2016–18 NISC Management Plan Report Card*, https://www.doi.gov/invasivespecies/2016-2018-nisc-management-plan-report-card; "Invasive Species Definition Clarification and Guidance," National Invasive Species Information

Center, 2006, https://www.invasivespeciesinfo.gov/invasive
-species-definition-clarification-and-guidance.

50. Mark Davis et al., "Don't Judge Species on Their Origins." *Nature* 474, no. 7350 (2011): 153.

51. "Cats and Birds," American Bird Conservancy, September 25, 2020, https://abcbirds.org/program/cats-indoors/cats-and-birds.

52. Note that in Australia, conservationists are indeed engaging in extensive eradication efforts against stray cats. See "Feral Cats—DCCEEW," December 20, 2024, https://www.dcceew.gov.au/environment/invasive-species/feral-animals-australia/feral-cats.

53. Jason M. Gleditsch and Tomás A. Carlo, "Fruit Quantity of Invasive Shrubs Predicts the Abundance of Common Native Avian Frugivores in Central Pennsylvania," *Diversity and Distributions* 17, no. 2 (December 29, 2010): 244–53.

54. Gleditsch and Carlo, "Fruit Quantity of Invasive Shrubs."

55. Quoted in Katrina Voss, "Invasive Plants Can Create Positive Ecological Change," Penn State, Eberly College of Science, September 1, 2020, https://science.psu.edu/news/invasive-plants-can-create-positive-ecological-change.

56. Voss, "Invasive Plants Can Create."

57. Quoted in Shiva, *Monocultures of the Mind*, 24. While the term "weed" is not necessarily always synonymous with the term "invasive species," invasive species are often called "weeds" and vice versa. And, of course, both weeds and invasive species will more often than not meet the same fate: merciless eradication efforts.

58. Vandana Shiva, "BBC - Radio 4 - Reith Lectures 2000 - Respect for the Earth," *BBC*, accessed November 3, 2025, https://www.bbc.co.uk/radio4/reith2000/lecture5.shtml..

59. A. Scheu et al., "The Genetic Prehistory of Domesticated Cattle from Their Origin to the Spread Across Europe," *BMC Genetics* 16 (2015): 54, https://doi.org/10.1186/s12863-015-0203-2.

60. Jacob Lipson, Travis Reynolds, and C. Leigh Anderson, "Environmental Implications of Livestock: Cattle," Evans School Policy Analysis and Research Group, July 31, 2011, https://epar.evans.uw.edu/environmental-implications-of-livestock-cattle.

61. Devatha P. Nair, "Cattle Ranching Is Actually Terrible for Biodiversity." *Sentient Media*, November 7, 2022, https://sentientmedia.org/cattle-ranching-terrible-for-biodiversity.

62. "Larkspur (Delphinium spp.)," U. S. Department of Agriculture, Agricultural Research Service, accessed July 22, 2025, https://www.ars.usda.gov/pacific-west-area/logan-ut/poisonous-plant-research/docs/larkspur-delphinium-spp.

63. Wulf Wilde, "Fact Check: How Bad Is Eating Meat for the Climate?," *DW (Deutsche Welle)*, October 30, 2022, https://www.dw.com/en/fact-check-is-eating-meat-bad-for-the-environment/a-63595148.

64. Bennett Rosenberg and Jared Hayes, "USDA Livestock Subsidies Top $59 Billion," Environmental Working Group, October 28, 2024, https://www.ewg.org/news-insights/news/2023/08/usda-livestock-subsidies-top-59-billion.

65. Shiva, *Monocultures of the Mind*, 27.

66. T. J. Demos, *Decolonizing Nature: Contemporary Art and the Politics of Ecology* (London: Sternberg Press, 2016), 150–51.

67. In South River Forest, the alleged objectionable invasive species include loblolly pine, box elder, Chinese privet, Callery pear, tree of heaven, chinaberry, and kudzu. However, loblolly pine and box elder are actually native to Atlanta.

68. Weelaunee was the name the Muscogee Cree people gave to the nearby South River.

69. Defend the Atlanta Forest, accessed July 22, 2025, https://defendtheatlantaforest.org.

70. Anna Kook, "Police vs. Atlanta: The Battle over Cop City," *Al Jazeera*, YouTube, 2022, https://www.youtube.com/watch?v=fPMW47tCA1g.

71. "People and Invasive Species," National Geographic, accessed July 22, 2025, 2023, https://education.nationalgeographic.org/resource/people-and-invasive-species.

72. *Climate Change Indicators in the United States*, 4th ed. (US Environmental Protection Agency, 2016), https://www.epa.gov/sites/default/files/2016-08/documents/climate_indicators_2016.pdf.

73. Brendon Larson, "Reweaving Narratives About Humans and Invasive Species," *Études rurales* 185 (2010), accessed March 31, 2018, http://journals.openedition.org/etudesrurales/9018.

74. David Pimentel, Rodolfo Zuniga, and Doug Morrison, "Update on the Environmental and Economic Costs Associated with Alien-Invasive Species in the United States," *Ecological Economics* 52, no. 3 (December 30, 2004): 273–88; Martina Igini, "Invasive Species Cost Global Economy $423bn Each Year, Threaten Ecosystems and Food Security: Report," Earth.org, September 5, 2023, https://earth.org/invasive-species-cost-global-economy-423bn-each-year-threaten-ecosystems-and-food-security-report.

75. Born in 1992 in Edward Wilson's book *The Diversity of Life*, this claim has held fast for years.

76. M. Chew, "Ecologists, Environmentalists, Experts, and the Invasion of the 'Second Greatest Threat,'" *Environment and Society Portal*,

accessed September 12, 2024, https://www.environmentandsociety
.org/mml/ecologists-environmentalists-experts-and-invasion-second
-greatest-threat.

77. Quoted in Chew, "Ecologists, Environmentalists, Experts."

78. Mark Davis, "Researching Invasive Species 50 Years after Elton: A Cautionary Tale," in *Fifty Years of Invasion Ecology*, ed. David Richardson (Chichester, UK: Wiley-Blackwell, 2011), 273.

79. D. S. Wilcove et al., "Quantifying Threats to Imperiled Species in the United States." *BioScience* 48, no. 8 (1998): 607–15.

80. J. Gurevitch and D. Padilla, "Are Invasive Species a Major Cause of Extinctions?," *Trends in Ecology and Evolution* 19, no. 9 (July 23, 2004): 470–74, https://doi.org/10.1016/j.tree.2004.07.005.

81. Dov F. Sax et al., "Ecological and Evolutionary Insights from Species Invasions," *Trends in Ecology and Evolution* 22, no. 9 (July 21, 2007): 467, https://doi.org/10.1016/j.tree.2007.06.009.

82. Quoted in Brian C. Chaffin et al., "Biological Invasions, Ecological Resilience and Adaptive Governance," *Journal of Environmental Management* 183, pt. 2 (December 1, 2016): 399–407, https://doi.org/10.1016/j.jenvman.2016.04.040.

83. Chaffin et al., "Biological Invasions, Ecological Resilience and Adaptive Governance."

84. Some do argue that biological invasions are the single leading cause of global extinctions. See Céline Bellard, Clara Marino, and Franck Courchamp, "Ranking Threats to Biodiversity and Why It Doesn't Matter," *Nature Communications* 13, no. 1 (May 16, 2022), https://doi.org/10.1038/s41467-022-30339-y.

85. Jennifer McDermott, Bernard Condon, and Michael Biesecker, "Bare Electrical Wire, Leaning Utility Poles Seen as Possible Cause of Deadly Maui Fires," PBS, August 27, 2023, https://www.pbs.org/newshour/nation/bare-electrical-wire-leaning-utility-poles-seen-as-possible-cause-of-deadly-maui-fires.

86. Shi En Kim, "How Swaths of Invasive Grass Made Maui's Fires So Devastating," *Smithsonian Magazine*, August 15, 2023, https://www.smithsonianmag.com/smart-news/how-swaths-of-invasive-grass-made-mauis-fires-so-devastating-180982729.

87. "Non-native Mountain Goats," Olympic Park Advocates, accessed July 22, 2025, https://olympicparkadvocates.org/non-native-mountain-goats.

88. Monoculture forests are far more prone to forest fires than more biodiverse forests. "Logging Impacts," Sierra Forest Legacy, accessed July 22, 2025, https://www.sierraforestlegacy.org/FC_FireForestEcology/FFE_LoggingImpacts.php.

89. "Fracaso: NAFTA's Disproportionate Damage to U. S. Latino and Mexican Working People," Public Citizen, October 16, 2020, https://www.citizen.org/article/fracaso-naftas-disproportionate-damage-to-u-s-latino-and-mexican-working-people.

90. "Immigrants and the Economy," American Civil Liberties Union, March 12, 2002, https://www.aclu.org/documents/immigrants-and-economy.

91. Katharine Abraham and Melissa Kearney, *Explaining the Decline in the U. S. Employment-to-Population Ratio: A Review of the Evidence* (Cambridge, MA: National Bureau of Economic Research, 2018), https://doi.org/10.3386/w24333.

92. Brian C. Chaffin et al., "Biological Invasions, Ecological Resilience and Adaptive Governance," *Journal of Environmental Management* 183 (July 3, 2016): 399–407, https://doi.org/10.1016/j.jenvman.2016.04.040.

93. Chaffin et al., "Biological Invasions, Ecological Resilience and Adaptive Governance."

Chapter 4: A New Story for Invasive Species

1. Lisa Newton and Catherine Dillingham, *Watersheds 3: Ten Cases in Environmental Ethics* (Belmont, CA: Wadsworth Publishing, 2001); Newton and Dillingham, *Watersheds 3*, 160.

2. "Eric Forsman Oral History Interview," Oregon State University, Special Collections and Archives Research Center, accessed July 23, 2025, https://scarc.library.oregonstate.edu/omeka/items/show/34846.

3. William Dietrich, *The Final Forest* (Seattle: University of Washington Press, 1992).

4. Mark Bonnett and Kurt Zimmerman, "Politics and Preservation: The Endangered Species Act and the Northern Spotted Owl," *Ecology Law Quarterly* 18, no. 1 (1991), 105.

5. Quoted in Richard Kluger, *The Bitter Waters of Medicine Creek: A Tragic Clash Between White and Native America* (New York: A. A. Knopf, 2011), 118.

6. Edwin Van Sickle, *They Tried to Cut It All: Grays Harbor—Turbulent Years of Greed and Greatness* (Seattle: Pacific Search Press, 1981). Note the genocidal and ecocidal implications in the word "develop."

7. Newton and Dillingham, *Watersheds 3: Ten Cases in Environmental Ethics*.

8. Kluger, *The Bitter Waters of Medicine Creek*, 374. Ellipses in original.

9. Newton and Dillingham, *Watersheds 3: Ten Cases in Environmental Ethics*.

10. Ernest G. Niemi et al., *The Sky Did Not Fall: The Pacific Northwest's Response to Logging Reductions* (Eugene, OR: ECONorthwest, 1999).

11. Elizabeth Shogren, "To Save Threatened Owl, Another Species Is Shot," NPR, January 15, 2014, https://www.npr.org/2014/01/15/262735123/to-save-threatened-owl-another-species-is-shot.

12. Kate St. John, "They Want to Cull Barred Owls Again," *Outside My Window*, November 30, 2023, https://www.birdsoutsidemywindow.org/2023/11/30/they-want-to-cull-barred-owls-again. This experiment was only remarkably successful in three out of five of the study sites. Despite the study claiming that the experiment was "consistently positive" in all five areas, the same study shows that two of the study sites indicate little to no difference in spotted owl population after hundreds of barred owls were killed.

13. John A. Wiens, *Ecological Challenges and Conservation Conundrums: Essays and Reflections for a Changing World* (Hoboken, NJ: John Wiley and Sons, 2016); "Barred Owl Life History," All About Birds, Cornell Lab of Ornithology, https://www.allaboutbirds.org/guide/Barred_Owl/lifehistory.

14. These are the two leading theories on how barred owls spread across the continent—via the Great Plains or the Canadian boreal forests. It seems most likely that the barred owls spread via both these routes, which is why I lumped the theories together. However, it is worth noting that the second theory, the Canadian boreal forest theory, is far less widely discussed than the first. One researcher has argued that the second theory may actually point to the barred owl as not being invasive at all. They point out that the boreal forests of Canada are over ten thousand years old and may have served as a natural route for the barred owls. This does not consider the settler suppression of Indigenous-controlled fires in the boreal forests as a possible reason for barred owl expansion, but, as these are all theories, it is worth considering the possibility that the boreal forest grew close enough to the eastern barred owl habitat to allow for a natural route. This researcher thoroughly explores the story of barred owls and points out the flaws in the plan to eradicate barred owls, and their article is well worth unpacking. See "For US Fish and Wildlife Service 'Management' Means Killing," Conservation Sense and Nonsense, January 1, 2024, https://milliontrees.me/2024/01.

15. "Barred Owl Letter," Animal Wellness Action, 2024, https://animalwellnessaction.org/wp-content/uploads/2024/11/Barred-Owl-letter-206.pdf.

16. "A Forest Stewardship Company," Green Diamond, 2025, https://www.greendiamond.com.

17. Natalie Jacewicz, "Humans Join Owl War in California," *Marin Independent Journal*, April 16, 2016, https://www.poconorecord.com/story/news/environment/2016/04/16/humans-join-owl-war-in/31668848007.

18. *3rd Annual Report*, submitted to the United States Fish and Wildlife Service and the California Department of Fish and Wildlife (Green Diamond Resource Company, 2023), https://www.greendiamond.com/downloads/2022_Forest_Habitat_Conservation_Plan_Annual_Report.pdf.

19. *Northern Spotted Owl Annual Report, 2023* (Humboldt Redwood Company, 2024), https://www.hrcllc.com/sites/default/files/inline-files/2023%20Northern%20Spotted%20Owl%20Annual%20Report.pdf; *3rd Annual Report* (Green Diamond Resource Company).

20. *Third Annual Report* (Green Diamond Resource Company). Note that "take" in this case means "to harass, harm, pursue, hunt, shoot, wound, kill, trap, capture, or collect, or to attempt to engage in any such conduct."

21. "Fish and Wildlife Service Gives Green Diamond New Permit to Take Spotted Owls," EPIC (Environmental Protection Information Center), August 5, 2019, https://www.wildcalifornia.org/post/fish-and-wildlife-service-gives-green-diamond-new-permit-to-take-spotted-owls; *3rd Annual Report* (Green Diamond Resource Company).

22. *Northern Spotted Owl Annual Report, 2023* (Humboldt Redwood Company).

23. *Final Barred Owl Management Strategy* (US Fish and Wildlife Service, 2024), https://www.fws.gov/sites/default/files/documents/2024-08/final-barred-owl-management-strategy-2024_508.pdf, 8.

24. Monica Bond, "Logging, Not Wildfire, Is Most Likely Driving Northern Spotted Owl Decline," *Phys.org*, October 25, 2022, https://phys.org/news/2022-10-wildfire-northern-owl-decline.html.

25. Throughout these documents, it is not only jargon that confuses but also plastic words. For example, these documents reference both a "Final Barred Owl Management Plan" as well as a "Spotted Owl Management Plan." Nearly identical in name, these plans are complete opposites in tactic. Unless one is well versed in invasive ecology, one will not necessarily realize at first glance that one instance of "management" means to save an owl while the other means to kill one.

26. Clare M. Schneider, "A Government Proposal to Kill a Half-Million

Owls Sparks Controversy," NPR, April 1, 2024, https://www.npr.org/2024/04/01/1241874707/california-spotted-owl-barred-owl-usfws.

27. It is worth noting that logging in the Pacific Northwest is not economically substantive—it doesn't actually turn a profit. However, because so much of the Pacific Northwest economy relies on logging and the timber industry, federal and state government subsidizes the net loss of profit. For example, the total wages and salaries in Washington for all industries is $352 billion per year, while forestry wages are $5.6 billion. The total annual Washington tax revenue for all industries and sources is $38 billion, while forestry tax revenue is $300 million. The timber industry generates a pathetic amount of tax revenue for the economy of Washington State. However, what the forestry wages and revenues don't account for are externalized costs, such as the long-term destruction of nature and the cost of maintaining timber export terminals and federal and state road infrastructure. "According to the [USFS], nationwide backlog maintenance projects top $1.2 billion, and funds available for road maintenance each year are only about 15 percent of what is needed to fully maintain the current road system." See "Sustainable Forest Roads," Conservation Northwest—Protecting, Connecting and Restoring Wildlands and Wildlife, May 14, 2019, https://conservationnw.org/our-work/wildlands/sustainable-forest-roads. The Forest Service facilitates the destruction of all these forests when it doesn't even generate enough revenue to maintain the timber roads. Additionally, agencies claim deforestation has been slowed or reversed because they fudge their numbers by counting monoculture plantations as forests. Meanwhile, they are branded with misleading sustainable forestry certifications. Today's forest industry is another example of a hollow, obsolete, and dead practice propped up by capital for capital.

28. Andrew S. MacDougall and Roy Turkington, "Are Invasive Species the Drivers or Passengers of Change in Degraded Ecosystems?," *Ecology* 86, no. 1 (January 1, 2005): 42–55, https://doi.org/10.1890/04-0669; Jonathan T. Bauer, "Invasive Species: 'Back-seat Drivers' of Ecosystem Change?," *Biological Invasions* 14, no. 7 (January 2, 2012): 1295–1304.

29. Stuart B. Weiss, "Cars, Cows, and Checkerspot Butterflies: Nitrogen Deposition and Management of Nutrient-Poor Grasslands for a Threatened Species," *Conservation Biology* 13, no. 6 (December 1999): 1476–86, https://escholarship.org/uc/item/81t6n5fb.

30. Susan E. Meyer et al., "Invasive Species Response to Natural and Anthropogenic Disturbance," in *Invasive Species in Forests and Rangelands of the United States*, ed. Therese M. Poland

et al. (Cham, Switzerland: Springer, 2021), 85–110, https://doi.org/10.1007/978-3-030-45367-1_5.

31. Andrew Carter and Lindsay Rosa, "Five Threats to Biodiversity and How We Can Counter Them," *Defenders of Wildlife*, July 24, 2023, https://defenders.org/blog/2023/07/5-threats-biodiversity-and-how-we-can-counter-them.

32. Brendon Larson, "The War of the Roses: Demilitarizing Invasion Biology," *Frontiers in Ecology and the Environment* 3, no. 9 (2005): 495.

33. Thomas Michael Bach and Brendon M. H. Larson, "Speaking About Weeds: Indigenous Elders' Metaphors for Invasive Species and Their Management," *Environmental Values* vol. 26 (Winwick, UK: The White Horse Press, 2017), https://doi.org/10.3197/0963271 17X15002190708119. It is also worth noting that one Elder said that without the weeds growing in the desert "[there would be] nothing out there on cattle country. . . . These plants protect country . . . without them, nothing, just dirt."

34. "Overview," Gaiatheory.org, https://web.archive.org/web/2019 0325153116/http://www.gaiatheory.org/overview.

35. Joanna Macy, *World as Lover, World as Self: A Guide to Living Fully in Turbulent Times* (Berkeley, CA: Parallax, 1991).

36. "DNR's 2022–25 Strategic Plan," Washington State Department of Natural Resources, https://www.dnr.wa.gov/strategicplan.

37. "Forest and Trust Lands," Washington State Department of Natural Resources, https://www.dnr.wa.gov/managed-lands/forest-and-trust-lands.

38. Hilary Franz et al., "Annual Report 2023," 2023, https://www.dnr.wa.gov/publications/em_annual_report_2023.pdf.

39. Jan D. Matthews, "An Introduction to the Situationists" (2005), Anarchist Library, https://theanarchistlibrary.org/library/jan-d-matthews-an-introduction-to-the-situationists.

40. Guy Debord, *The Society of the Spectacle* (1967), Anarchist Library, https://theanarchistlibrary.org/library/guy-debord-the-society-of-the-spectacle.

41. "Citizen Science," *National Geographic*, accessed September 9, 2025, https://education.nationalgeographic.org/resource/citizen-science-article.

42. "About CitizenScience.gov," accessed July 23, 2025, https://www.citizenscience.gov/about.

43. Poerksen, *Plastic Words*.

44. Karel Mulder, Didac Ferrer, and Harro Van Lente, *What Is Sustainable Technology?: Perceptions, Paradoxes, and Possibilities* (New York: Routledge, 2017), 236–37.

45. Laura Ferguson, "The Extinction Crisis," *Tufts Now*, May 21, 2019, https://now.tufts.edu/2019/05/21/extinction-crisis.

46. Tammana Begum, "What Is Mass Extinction and Are We Facing a Sixth One?," Natural History Museum, May 19, 2021, https://www.nhm.ac.uk/discover/what-is-mass-extinction-and-are-we-facing-a-sixth-one.html.

47. Martina Igini, "6 Biodiversity Loss Statistics That Will Blow Your Mind," Earth.org, August 19, 2022, https://earth.org/biodiversity-loss-statistics.

Chapter 5: Solutions

1. Karl Marx, *Capital: A Critique of Political Economy*, vol. 1, trans. Ben Fowkes (New York: Penguin Books, 1976), 637–38.

2. Around this same time, fear around overpopulation and general racist perspectives led to the United States organizing eugenic practices in Puerto Rico, including the forced sterilization of women on the island.

3. Fred Pearce, *The New Wild: Why Endangered Species Will Be Nature's Salvation* (Boston: Beacon Press, 2015), 104.

4. Pearce, *The New Wild*, 109–10. Yes, this math is wrong. I can only assume he mean the remaining 33 percent, not the remaining 67 percent.

5. Pearce, *The New Wild*, 105.

6. Pearce, *The New Wild*, 110.

7. "How Feedback Loops Are Making the Climate Crisis Worse," Climate Reality Project, January 7, 2020, https://www.climaterealityproject.org/blog/how-feedback-loops-are-making-climate-crisis-worse.

8. "How Feedback Loops Are Making the Climate Crisis Worse."

9. Tao Orion, *Beyond the War on Invasive Species: A Permaculture Approach to Ecosystem Restoration* (White River Junction, VT: Chelsea Green, 2015), 103.

10. "Indigenous Knowledge Is Crucial in the Fight Against Climate Change—Here's Why," Climate Promise, United Nations Development Programme, July 31, 2024, https://climatepromise.undp.org/news-and-stories/indigenous-knowledge-crucial-fight-against-climate-change-heres-why.

11. Delilah Friedler, "California's Wildfire Policy Totally Backfired. Native Communities Know How to Fix It.," *Mother Jones*, November 11, 2019, https://www.motherjones.com/environment/2019/11/californias-wildfire-controlled-prescribed-burns-native-americans.

12. Kyle Whyte, "Indigenous Climate Change Studies: Indigenizing Futures, Decolonizing the Anthropocene," *English Language Notes*

55, no. 1–2 (March 1, 2017): 157, https://doi.org/10.1215/00138282-55.1
-2.153.

13. "Indigenous Knowledges and Climate Change," Climate Atlas of Canada, accessed September 7, 2025, https://climateatlas.ca/indigenous-knowledges-and-climate-change.

14. Whyte, "Indigenous Climate Change Studies: Indigenizing Futures, Decolonizing the Anthropocene," 157.

15. Whyte, "Indigenous Climate Change Studies: Indigenizing Futures, Decolonizing the Anthropocene," 154.

16. Nick J. Fox, "Green Capitalism, Climate Change and the Technological Fix: A More-than-human Assessment," *The Sociological Review* 71, no. 5 (September 23, 2022): 1115–34, https://doi.org/10.1177/00380261221121232.

17. "Greta Thunberg's 'blah blah blah' speech, Milan 2021," Carbon Independent, https://www.carbonindependent.org/119.html.

18. Jasper Bernes, "Between the Devil and the Green New Deal," *Commune*, May 26, 2019, https://communemag.com/between-the-devil-and-the-green-new-deal.

19. "Filthy Future: Wind and Solar's Toxic Waste Legacy Problem for the Next Generation," Stop These Things, https://stopthesethings.com/2021/11/25/filthy-future-wind-solars-toxic-waste-legacy-problem-for-next-generation.

20. Neil Blakemore, "Non-Profits Can Not Liberate Us, but Mutual Aid and Solidarity Can Be a Pathway Forward," *Medium*, July 27, 2023, https://medium.com/@neilblakemore/non-profits-can-not-liberate-us-but-mutual-aid-and-solidarity-can-be-a-pathway-forward-6f4197cecacd.

21. Gilles Dauvé, *Pommes De Terre Contre Gratte-ciel: Critique de L'Écologie Politique*, 2024. My translation.

22. Audra Mitchell, "Decolonising the Anthropocene," 2015, https://worldlyir.wordpress.com/2015/03/17/decolonising-the-anthropocene, as cited in Whyte, "Indigenous Climate Change Studies: Indigenizing Futures, Decolonizing the Anthropocene," 157.

23. Tammy Gan, "Welcome to the ZAD, a Site of Reclaiming and Resistance," *Shado Magazine*, May 3, 2022, https://shado-mag.com/articles/act/welcome-to-the-zad-a-site-of-reclaiming-and-resistance.

24. "Reflections on the ZAD: Another History," *CrimethInc.*, 2019, https://crimethinc.com/2019/04/23/reflections-on-the-zad-looking-back-a-year-after-the-evictions.

25. "Reflections on the ZAD." The entire story of the ZAD is complex, particularly at the end of its struggle when a sanction of the ZAD signed a contract with the state to keep the land safe from developers. For many, this compromise with the state was a capitulation. Despite

the complexity, it is arguably an example of people defending nature from the prophesized corruption and destruction of capital. Thousands of acres of what would have become barren land and polluting airport were preserved and remain stewarded by the people who love it and live with it.

26. Gan, "Welcome to the ZAD, a Site of Reclaiming and Resistance."

27. "Earth Uprisings – A Story," lessoulevementsdelaterre.org, 2023, https://lessoulevementsdelaterre.org/blog/earth-uprisings-une-histoire.

28. "Treaties Still Matter: The Dakota Access Pipeline," National Museum of the American Indian, https://americanindian.si.edu/nk360/plains-treaties/dapl. Ellipses in source.

29. Nick Estes, *Our History Is the Future: Standing Rock Versus the Dakota Access Pipeline and the Long Tradition of Indigenous Resistance* (London: Verso, 2019), 36, 17.

30. Estes, *Our History Is the Future*, 24.

31. Shelia Hu, "The Dakota Access Pipeline: What You Need to Know," Natural Resources Defense Counsil, June 12, 2024, https://www.nrdc.org/stories/dakota-access-pipeline-what-you-need-know.

32. "The #NoDAPL Movement Was Powerful, Factual, and Indigenous-Led. Lawsuit Lies Can't Change That." Center for Constitutional Rights, February 21, 2018, https://ccrjustice.org/home/blog/2018/02/21/nodapl-movement-was-powerful-factual-and-indigenous-led-lawsuit-lies-can-t.

33. Amanda Stephenson, "The Keystone Pipeline's History of Spills." *Reuters*, April 9, 2025, https://www.reuters.com/business/environment/keystone-pipelines-history-spills-2025-04-09.

34. Estes, *Our History Is the Future*, 177.

35. "Prise de Terre(s)," Nadir.org, 2019, https://zad.nadir.org/spip.php?article6658.

36. Estes, *Our History Is the Future*, 22.

37. Other notable environmental movements include the Chipko movement of the Himalayas in the 1970s; the Green Belt movement in Kenya (also in the 1970s); the Landless Worker Movement in Brazil in the 1980s; the Zapatistas in Chiapas, Mexico, which started in the 1980s; Earth First!, which began in the United States in the 1980s; and the Earth Liberation Front, which began in England in the 1990s. There have been and are many more.

38. "Prise de Terre(s)."

Chapter 6: From the Old World, We Know

1. David Hinton, *Wild Mind, Wild Earth: Our Place in the Sixth Extinction* (Boulder, CO: Shambhala, 2022), 36.

2. Patty Krawec, *Becoming Kin: An Indigenous Call to Unforgetting the Past* (Minneapolis: Broadleaf Books, 2022), 8.

3. Estes, *Our History Is the Future*, 24.

4. Quoted in Johnny Langenheim, "Natural Custodians: Indigenous Lessons in Reconnecting With Nature," *Environment*, March 1, 2024, https://www.nationalgeographic.com/environment/article/paid-content -natural-custodians-indigenous-lessons-in-reconnecting-with-nature.

5. "Harmony with Nature? Not in Ancient America," *Deseret News*, March 19, 1995.

6. Karl Marx, *Economic and Philosophic Manuscripts of 1844*, trans. Martin Milligan (Moscow: Progress Publishers, 1959), 31, https://www.marxists .org/archive/marx/works/download/pdf/Economic-Philosophic -Manuscripts-1844.pdf.

7. Marx, *Capital*, vol. 1, 637.

8. Marx, *Economic and Philosophic Manuscripts of 1844*, 28–35.

9. Marx, *Capital*, vol. 1, 638.

10. Krawec, *Becoming Kin*, 167.

11. "An Ethics of Wild Mind: An Interview with David Hinton," *Emergence Magazine*, February 7, 2023, https://emergencemagazine.org/ conversation/an-ethics-of-wild-mind.

12. Estes, *Our History Is the Future*, 25.

13. Fiona Harvey, "Major Climate Changes Inevitable and Irreversible— IPCC's Starkest Warning Yet," *Guardian*, August 9, 2021.

14. "Prise de Terre(s)."

15. Priscilla M. Wehi et al., "Contribution of Indigenous Peoples' Understandings and Relational Frameworks to Invasive Alien Species Management," *People and Nature* 5, no. 5 (July 17, 2023): 1403–14, https://doi.org/10.1002/pan3.10508.

16. Nicholas J. Reo and Laura A. Ogden, "Anishnaabe Aki: An Indigenous Perspective on the Global Threat of Invasive Species," *Sustainability Science* 13, no. 5 (May 4, 2018): 1443–52, https://doi.org/10.1007/s11625 -018-0571-4, 6.

17. Reo and Ogden, "Anishnaabe Aki: An Indigenous Perspective on the Global Threat of Invasive Species," 2, 6.

18. Reo and Ogden, "Anishnaabe Aki," 7.

19. "October 22, 1928: Principles and Ideals of the United States Government," Miller Center, October 20, 2016, https://miller center.org/the-presidency/presidential-speeches/october-22-1928 -principles-and-ideals-united-states-government.

20. Gilles Dauvé, *From Crisis to Communisation* (Oakland: PM Press, 2019), 48.

21. Generally speaking, historians and anthropologists argue that hunter-

gathers were most common early humans (millions of years ago) up until 10,000 years ago, which coincides with the rise of agriculture. While these timescales are not exact, they still indicate that egalitarianism among ancient peoples was relatively common for possibly millions of years—an imprecise but incredible track record for equality among human beings.

22. Deborah Rogers, "Inequality: Why Egalitarian Societies Died Out," *New Scientist*, July 25, 2012, https://www.newscientist.com/article/dn22071-inequality-why-egalitarian-societies-died-out.

23. G. Balthazar Gras, "Meaningful Action: Recomposing Labor and Value with Practices and Imaginaries on the Zad of the Notre-Dame-de-Landes," *Journal of Intersectional Social Justice*, Winter 2005, https://jisj.pubpub.org/pub/lcxrux9t/release/1.

24. "Acorn Community Farm," Foundation for Intentional Community, https://www.ic.org/directory/acorn-community-farm.

25. "Cambia No Longer," Cambia Community, https://cambiacommunity.weebly.com.

26. "Can Inequality Be Blamed on the Agricultural Revolution?," World Economic Forum, October 25, 2018, https://www.weforum.org/stories/2018/10/how-the-agricultural-revolution-made-us-inequal.

27. "Nearly 9 in 10 People Globally Want a More Sustainable and Equitable World Post COVID-19," World Economic Forum, September 16, 2020, https://www.weforum.org/press/2020/09/nearly-9-in-10-people-globally-want-a-more-sustainable-and-equitable-world-post-covid-19.

28. Marx, *Economic and Philosophic Manuscripts of 1844*, 29, 31.

29. Judy Cox, "An Introduction to Marx's Theory of Alienation," *International Socialism*, July 1988, Marxists Internet Archive, https://www.marxists.org/history/etol/newspape/isj2/1998/isj2-079/cox.htm.

30. Rogers, "Inequality."

31. Chris Byron, "Essence and Alienation: Marx's Theory of Human Nature," *Science and Society* 80, no. 3 (July 2016): 383, https://philarchive.org/archive/BYREAA-2.

32. Safa Motesharrei, Jorge Rivas, and Eugenia Kalnay, "Human and Nature Dynamics (HANDY): Modeling Inequality and Use of Resources in the Collapse or Sustainability of Societies," *Ecological Economics* 101 (May 2014): 90–102, https://doi.org/10.1016/j.ecolecon.2014.02.014.

33. Rebecca Solnit, *A Paradise Built in Hell: The Extraordinary Communities That Arise in Disaster*, rev. ed. (New York: Penguin Books, 2020), 4–5.

34. Solnit, *Paradise Built in Hell*, 8.

35. "Natural Disasters Creating Oppressive Governments, New Study Finds," Institute of Sustainability and Environmental Profession-

als, accessed July 25, 2025, https://www.isepglobal.org/articles/natural-disasters-creating-oppressive-governments-new-study-finds.

36. Solnit, *Paradise Built in Hell*, 49–50.

37. "Natural Disasters Creating Oppressive Governments, New Study Finds," press release, Response Source, August 7, 2019, https://pressreleases.responsesource.com/news/98191/natural-disasters-are-creating-oppressive-governments.

38. Quoted in "Natural Disasters Are Creating Oppressive Governments."

39. Solnit, *Paradise Built in Hell*, 10.

40. Blakemore, "Non-Profits Can Not Liberate Us."

Conclusion: The Colonization of the English Language

1. Vinciane Despret, "Afterword: It Is an Entire World That Has Disappeared," in *Extinction Studies: Stories of Time, Death, and Generations*, ed. Deborah Bird Rose, Thom Van Dooren, and Matthew Chrulew (New York: Columbia University Press, 2017), 217.

2. United Nations, "Biodiversity - Our Strongest Natural Defense Against Climate Change, https://www.un.org/en/climatechange/science/climate-issues/biodiversity. Accessed September 16, 2025.

3. "Words—Transcript," interview by Robert Krulwich and Jad Abumrad, 2010, https://radiolab.org/podcast/91725-words/transcript. Even adults will behave like rats under certain conditions that render language meaningless. The researchers had adults listening to a speech given the task of repeating the words of the speech as fast as they could. This essentially disabled their spatial awareness, and they too could not comprehend the idea of something being left of the blue wall.

4. SCOPUS is one of the biggest databases for peer-reviewed journals in the world. Huttner-Koros, "The Hidden Bias of Science's Universal Language."

5. Huttner-Koros, "The Hidden Bias of Science's Universal Language."

6. Quoted in Huttner-Koros, "The Hidden Bias of Science's Universal Language."

7. Huttner-Koros, "The Hidden Bias of Science's Universal Language."

8. Poerksen, *Plastic Words*, 2, 3.

9. Sachs, *Planet Dialectics*, 93.

10. Huttner-Koros, "The Hidden Bias of Science's Universal Language."

11. Sachs, *Planet Dialectics*, 93.

12. Sachs, *Planet Dialectics*, 94.

13. Shiva, *Monocultures of the Mind*, 11.

14. Shiva, *Monocultures of the Mind*, 12, 9, 10.

15. Shiva, *Monocultures of the Mind*, 9–10, 12.